Glendale College
Library

Against
Pollution and
Hunger

ALICE MARY HILTON (ed.)

Against Pollution and Hunger

Society for
Social Responsibility in Science

A HALSTED PRESS BOOK

John Wiley & Sons
New York – Toronto

Published in Europe and the United Kingdom by
Universitetsforlaget, Oslo, Norway

Published in the U.S.A., Canada and Latin America
by Halsted Press,
a Division of John Wiley & Sons Inc.,
New York

Library of Congress Cataloging in Publication Data
Main entry under title:

Against pollution and hunger.

"A Halsted Press book"
Papers presented at a conference, sponsored by the Society for Social Responsi-
bility in Science, held at the University of Trondheim
1. Pollution – Congresses. 2. Science – Social aspects – Congresses. I. Hilton, Alice
Mary, ed. II. Society for Social Responsibility in Science.
TD172.5.A34   363.6   73-17751

ISBN 0-470-39771-6

Printed in Norway by
Aktietrykkeriet i Trondhjem

To Maurice Strong

whose brilliant leadership
in the struggle for the
environment might yet
win us the battles
against pollution and hunger

v

Editor's Preface

It is natural — and quite proper — for the organizers of a major international scientific conference to ask themselves whether all the effort and money for the preparation of the conference was spent appropriately and in a worthwhile manner. Do the results of the conference, in other words, justify the energy spent by the organizers and the money that has been entrusted to them by the members of the co-sponsoring organizations?

Apart from the intangible effects of any scientific conference, such as the stimulation and cross-fertilization of the minds of participants, this conference passed some resolutions and made some recommendations to individuals and governments. Perhaps the most tangible result is this book. The reader, therefore, can judge for himself whether the conference was worth the effort.

As editor of the proceedings and president of the sponsoring society, I am not the impartial judge that is needed to weigh the results. I am, however, very familiar with the enormous effort that went into the preparation of the conference and I should like to express my sincere thanks to the many people whose cooperation and devotion made the conference and this book possible. Above all, the authors of the conference papers — the contributors to this book — deserve the most sincere thanks for sharing their work and ideas so selflessly.

Our thanks are due to our distinguished board of sponsors of the conference, to the Conference Chairman, Professor Victor Paschkis of Columbia University in New York, and to Professor Harald Wergeland and his committee at Trondheim University.

We deeply appreciate the financial assistance given by many members of the Society for Social Responsibility in Science and the generous contribution made by the Royal Norwegian Academy of Science. Without their help the conference would have been impossible.

Most of all, we are indebted to the University of Trondheim, the host institution of this conference, for its generous hospitality and for the personal efforts of the Rector, Dr. Gunnar Bøe, and the Dean, Dr. Kjell Carlsen, without whose wise counsel this conference could not have succeeded.

Many others gave generously of their time and energy. It is impossible to name them all. Space permits me to mention only a few: Mrs. Erna Odde and Mrs. Gunhild Herø, who, throughout the summer of preparation, put up so cheerfully with my (often undecipherable) English notes and transmitted them into impeccable Norwegian, French, German . . .

Particularly, I want to thank Professor Per Hemmer who quietly saw to it that everything worked, that everyone was happy, that ruffled tempers were smoothed before the ruffles had a chance to show, who drafted and redrafted hazy resolutions until they were clear, who did a thousand things no one even knows about to assure the success of the conference.

The papers presented at the conference show that scientists, like non-scientists, rarely agree on political measures that must be taken to cope with scientific facts about which there is little disagreement. Every participant agreed that there are serious problems in our world. Pollution *is* serious indeed. And so is hunger. Those are facts that can be demonstrated and measured and proved. But what is to be done about them is not so simple. For that is, of course, a political issue of major proportions. And here we did not find it so easy to agree. Many of us know quite well what *ought* to be done although we do not always agree precisely on that. But some of the conferees were more acutely aware of the vast chasm from the shore of 'what *ought* to be done' to the bank of 'what *can* be done.'

It is in the realm of 'can' that difficulties arise. It is difficult to identify the vast streams and powerblocks that are keeping the 'ought' from the 'can', for these powerblocks are so often shrouded in fog. Two of the papers point to the difficulties of enacting and enforcing national legislation and international agreements, respectively. It is all too easy to point to the 'rascals' — big industry or big labor or big government — as the powers that divide 'ought' from 'can'. But much bigger than the biggest of the 'rascals' is the vast majority of mankind with its many vaguely formulated desires and often conflicting wants. And only in the very broadest sense, in the most profound depth, and in the long run are the wants and desires visible as being basically the same for all human beings: life. Life, first of all. And the Good Life, if possible. And there's the rub! For what is the Good Life? To a university professor in the rich, industrialized nations? To the peasants in Southeast Asia? To the wretched in the favelas of Brazil? To the villagers of India and China? To the craftsmen of Africa?

Most human beings are too poor and hungry to worry about the Good Life. But we, in the privileged, industrially advanced nations, who have forgotten about hunger and pestilence generations ago, do worry about the Good Life — and not only the Good Life for ourselves. And because we worry, some among us deplore technology and industrialization and long for the blessed solitude of a beautiful wilderness of our own choosing. But that is only for the lucky ones, the affluent in the rich nations, who can afford the enormous luxury of making choices. Only they can dream of choosing between a house in the suburbs and a mountaintop! Then, who would not want the mountaintop?

But for the vast majority of mankind there are no such choices. For them there are no choices at all. Their desires are much simpler and more primitive. They just want to survive. They want to still their hunger pangs and their babies' cries for bread. They want simple food, a roof to keep them dry, and clothes to keep them warm. And only when these most basic needs have been met will they be able to think of such unheard-of luxuries as medicines to soothe their bodies and schools to excite their minds.

This is the immense and complex dilemma of our world: How to produce enough to feed the hungry; how to produce enough to clothe the ragged and to house the homeless. How to preserve the earth so that life is worth living. How to grow enough food so that nobody need ever go hungry again and yet not to deprive the affluent of the wilderness they cherish. For in a world in which human beings can thrive and civilization flourish, it is not enough to provide mere survival. Though that must come first and foremost.

Many of us worry about the excesses of consumption by the affluent in the rich nations. We deplore the waste. We detest the shoddiness. We loathe the nasty vulgarity of too many badly made cars, too much polluting detergent, too many over-flavored foods. We worry about the poor quality we have bought with our vast quantity. We worry also lest those who do not have enough quantity to worry about quality at all might learn nothing from their more affluent brothers, and that they too will sacrifice all chance for quality to their legitimate desire to have enough quantity.

That is a great danger. And only in the rich nations can we afford the luxury of worrying about it. One does not have the problem of choosing a life style until one's life is assured and comfortable and safe.

The wilderness and the natural beauty of the earth are, and must be cherished by those who can afford this wonderful luxury. If it were possible — even in the long run — to feed humanity at the expense of our rivers and our mountaintops, if we could house the homeless at the expense of the magnificence created by human genius — the Taj Mahal or the Cathedral of Chartres, a Bach cantata, Rembrandt's painting — we might consider it our duty to sacrifice the mountaintops and the treasures of civilization, even though we would have to exchange for human survival everything that makes human life worth living.

But it is not possible to make such choices. We'd soon run out of the wilderness and mountaintops and the beautiful monuments created by human genius — and the majority of mankind would still not have enough to eat.

The choice before us is not one of peaks for the few or food for all. That might often have been true in the past, simply because mankind did not know how to produce enough with the resources that had been developed. But now, for the first time in human history, it is, indeed, possible to produce enough for human beings to live, and still to preserve a world fit for human beings to live in.

Of course, three billion human beings cannot climb mountaintops or they would be as crowded as the favela-covered hills of Rio de Janeiro. But it is not really so important to have enough mountaintops to go around. There are few enough people who really *want* to make the effort to climb a mountain, few enough who long for the wilderness.

What the few want and desire is really not of the slightest socio-economic or political significance. But it is of the greatest moral and and pedagogic importance. The few set the tone. Their values are adopt-ed — sifted perhaps and often soiled, but adopted just the same — by the many when they can adopt them. Just look at fashions. Gold and diamonds for the few, and colored plastic for the many as soon as it can be mass-produced. That is why the few, who are privileged in means and education, have such a vast responsibility. They have many privi-leges, and above all is the privilege to make choices. For that privilege of choice, of being able to indulge their personal preference, they must accept the moral responsibility of setting the moral climate of civili-zation and of thinking before they invite all others to follow them whether the others can do so and whether they would be happy. Above all, the few must not be so selfish as to want to make the whole world into a playground for their particular hobbies. They, too, must learn to look at the world and see it as it is and work for what can be, and love the world that is and can be.

The mountaintop that is so beloved by the privileged few is but a precious oasis in the vast desert of poverty and want that can only be helped if enough is produced for all. If the few cherish the privilege of choice, they must not seek to preserve it by objecting to all technology and industrialization. For that would be to seek to preserve the privi-lege — and expand it — at the expense of the vast majority of human beings, those who are still starving and homeless, in this world.

The privileged do not have the right to be irresponsible and thought-less. Nor do they have the right to think that the great dilemma of our time can be solved by hoping for the disappearance of the many for whom more must be produced. Who is to do the disappearing? The 'surplus' population? The poor? And are they to have nothing to say about it? For surely nobody thinks that the refugees from Bangla Desh, the children in the favelas of South America, the refugees in Indo-China will oblige us, the rich, who like our wilderness better than our crowd-ed surburbs, by disappearing, by no longer geing born. They will not oblige us. Astonishing as it may be, they want to survive. They have nothing, the poor on this earth, but their sad lives. Sad lives, but to them precious. And the rich do not have the right to wish away the only possession of the poor: their lives. No, the privilege of being rich must be earned by more strenuous efforts — by finding the means to produce enough on this earth to feed all who are hungry and by doing so *without* allowing the earth to be spoiled for the enjoyment of all human beings — each in his own way, including the way of finding a wilderness that offers blessed solitude and peace.

That was the real purpose of the conference: to think together and share our knowledge and experience. To find a way against pollution *and* hunger. And to find the way that leads towards a world in which all human beings can live human lives, each in his own fashion, and, within reason, according to his own desires, and each allowing his neighbours the same privilege of free choice that he has every right to claim for himself.

In this framework, then, it would be impossible for me to claim great success for the conference — success, in the sense of having arrived at definitive conclusions of what *ought* to be done as well as what *can* be done. But it would have been unreasonable to expect that. The best that we had any reasonable hope for was that the problems would be identified, human desires explained, conflicts recognized, deep roots of the dilemma exposed, remedies suggested for careful exploration — a much more careful and intensive exploration than one could hope to at-tempt in a three-day conference.

A.M.H.

Contents

Acknowledgements

Since its birth, the Society for Social Responsibility in Science has been based on two premises: first, that every engineer and scientist is personally, morally, and socially responsible for the foreseeable consequences of his work; second, that science, medicine, and technology should be used only for constructive purposes.

Originally, emphasis was put on questions of armaments, war and peace. But soon it became clear that, if man avoids nuclear suicide, modern technology still places him on a deadend course. Responding to this rapidly developing crisis, SSRS has grown in scope and program. Our annual 1971 conference bears witness to the breadth of this growth. Over two hundred scientists and engineers came from many countries to share their concern and seek new solutions.

Papers and panels covered a wide spectrum. Starting from the physical and chemical aspects of pollution, the program dealt with health aspects, with the question of pollution in relation to the technically less developed countries, with education, and with the question of what a pollution-conscious technology requires in the way of life-style.

The conference was not all talk and discussion. A resolution was accepted for submission to the June 1972 UN Conference on the Human Environment in Stockholm. A call for immediate action reminds us all that knowledge must be translated into action now in order to be effective. Further, a defense fund was started to support scientists, who, because of principled actions for the defense of mankind against pollution, suffer prosecution and other difficulties.

We gratefully acknowledge our debt to the Royal Norwegian Academy of Science and to the University of Trondheim for co-sponsoring this Conference on International Pollution Control. Our deepest thanks go to Dr. Harald Wergeland, past President of the Academy, Professor of Theoretical Physics, and a member of the SSRS Council, for his enormous contribution of time and thought. We also thank the local committee and the City of Trondheim. None of us will forget the warmth and grace of the social occasions or the beauty in which they were set.

Scientists and engineers from many lands went home freshly aware of their awesome responsibility to warn mankind of dangers and to seek ways to make all life safe on this 'spaceship earth'. SSRS has pioneered in this concept. Let us now work harder and grow.

Victor Paschkis
Chairman, 1971 SSRS Conference

Introduction

RALPH NADER

The Responsibility of
the Professional

The major factor responsible for
increasing pollution since 1946 is not
the increased number of people, but
the intensified effect of economically
faulty technology on the environment.
(Barry Commoner)

Much has been said about the 'responsibility of professionals.' Lawyers
have obligations to their clients, but they also have sometimes conflicting
obligations to the legal system they serve and to the general citizenry
the legal system purports to protect. Doctors owe duties to their patients, but they also owe duties to medical science and to the general
public's health and well-being. Many engineers also have corporate employers or clients to whom they owe their loyalty, but they also are required with varying degrees of compulsion to honor codes of ethics,
which mandate that they never sacrifice the health and safety of others
to suit the whims of their clients. And so it goes with all professionals
whose professions have developed from a tradition of supplying independent judgement based on expertise and education to the problems and
purposes of their employers or clients.

Scientists are usually placed, or place themselves, in a slightly different category. They are bound to search for the truth, for the hard
facts, for the essential realities in their respective disciplines. They are
compelled to discover truth, and in a sense they view truth as their
'clients'. To say all this about scientists is to state but the first premise.
Without elaboration it says absolutely nothing about a scientist's obligations to his fellow man. Unfortunately, too many scientists are content
to rest their commitment to 'social responsibility' on this empty truism.
To do so is to embrace a particularly virulent form of *hubris* that has no
place among people who, to a substantial degree, hold the keys to the
future of mankind.

Other scientists go one step further. They seek to know the consequences of their work. Having discovered one or more 'truths', they
feel compelled, and happily so for the rest of the community, to ask:
'To what uses will my new knowledge be put, and how should I seek
to influence those uses?' This is the critical point at which even the

19

scientist who is most enlightened about social responsibility, or thinks he is, can easily fail to do anything other than satisfy his own conscience in some small way.

At this point the concerned scientist is in roughly the same position as any other professional with clients or an employer. He must, publicly or privately, choose between his allegiances, his duties, and his obligations to his immediate employer or even to some idealized notion of 'the truth' and his responsibilities to his fellow men or, more specifically, to those who will be affected by the uses to which his work will be put.

In 1967 the Study Committee on Ethics and Responsibilities of Scientists of the American Academy of Arts and Sciences under the direction of Anatol Rappoport sampled the opinions of Academy members of various ethical questions. Over 80 percent of the respondents 'agreed that devotion to science implies also a devotion to certain values, such as concern for human welfare, etc., and that a part of a scientist's responsibility is to help in promoting these values and not to help in promoting goals incompatible with them.' The survey revealed the same high level of agreement with the proposition that 'if they found evidence of dishonesty within their organization they would protest (at least) within the organization.' *And* between 70 and 80 per cent said they 'would publicize their opinion if they thought that a drug released for the market had not been sufficiently tested for safety.'

The survey also posed the following hypothetical situation:

> You find evidence that organs of the government withhold or suppress information (apparently for policy reasons) which you feel very strongly should be available to the public. You have this information and can rely on its source. Assume that, while the dissemination of the information is legal, it undercuts government policy or is embarrassing to the government.

Of the respondents, 52.5 per cent indicated they would disseminate such information under certain circumstances. Unfortunately, the situation was not drawn up to elicit responses for cases in which the scientist is employed by, or is an advisor to the particular agency suppressing the information. In the United States it is quite typical for large segments of the important leadership of a scientific discipline to be drawn formally into some kind of government advisory body. Some have argued most persuasively that this tends to co-opt dissenting views within the scientific establishment. And it is in this situation that the scientist's responsibilty to his fellow citizens is greatest.

There are signs that at least some scientists are moving toward transforming these opinions into behavioral facts — to give them institutional meaning and, above all, to put them into practice. The Society for Social Responsibility in Science (SSRS) is instrumental in this development.

There are generally two avenues along which scientists and technologists, whether academic or industrial, can influence decisions that have an impact on the lives and livelihoods of their fellow citizens. They can act through organizations like professional societies and issue-oriented associations, *or* they can act individually. These two possibilities are not mutually exclusive. Depending on the circumstances, it may be necessary and appropriate to act collectively or individually or both. Of course, there are a multitude of possible variations within each of these avenues. SSRS represents one form of the collective approach. So do recent efforts *within* some of the established scientific and professional societies.

The special obligation of any society of professionals — a society of engineers or scientists, no less than one of lawyers or doctors — is to be more than the sum of its parts. A society or an organization like SSRS must have more influence, conduct more significant activity, be more aggressive in the protection and advancement of the public interest, and more effectively advance professional responsibility and guarantee professional independence than any of its members, were they to act individually.

Too often, professional societies in general, and engineering and scientific societies in particular, are content to issue vague moral pronouncements about moral or professional responsibility. Societies frequently 'roll over and play dead' when truly critical issues are raised: What is the society doing to defend and extend the rights of members or would-be members to challenge clients or employers who seek to stifle dissent?

You may have heard of the case of A. Ernest Fitzgerald, a high official in the US Department of the Air Force who was fired (or, more precisely, had his position abolished) for relentlessly doing what he was hired to do: control defense contract or cost overruns. Mr. Fitzgerald 'blew the whistle' on the C5A (a giant transport airplane) cost overruns in a testimony before a US Senate Committee. Mr. Fitzgerald appealed to his professional society for support. He was a founding member of the organization and felt reasonably confident that they would take some action on his behalf. Their response: At a special meeting of the board convened to discuss his case, it was decided that from that day forth the society was to be a 'technical' and not a 'professional' society and that it therefore no longer had any jurisdiction over his case and others like it.

In reality, professional societies in general, particularly those in the engineering and scientific areas, have tended to be indentured to the particular corporations, industries, or government agencies that deal with their subject matter. The automotive and chemical industries, for example, can easily control the societies that deal with automotive science or chemistry.

The other sources of control are even more overt in the sense that the companies provide full expenses for whatever testing is done — for example, in engineering areas with company property. The topics of the symposia are determined by engineering societies on company missions, not on their own professional missions. The fact that there are crises in such areas as pollution, pesticides, auto safety, and so forth indicates that there is quite a spread between what the companies are doing by way of allocating resources or backing up scientific and technical personnel and what the professional missions of these bodies of knowledge are.

A second reality of professional societies is that they do not defend the whistle-blower. I think every individual working for an organization — whether he is a professional or a technican or a floor-sweeper — has to keep in mind where the line is going to be drawn beyond which his allegiance to society supersedes his allegiance to the organization. This has to be so. There must be some sort of inner, initial determination that the individual employee will only go so far in obeying the dictates of the organization and beyond that will have to blow the whistle and achieve a resolution of his own conscience by communicating his concern and his information to outside authorities.

This is, basically, the principle established at the Nuremberg Trials: You can only go so far in excusing your performance by saying that you are just taking orders; there is a limit, for instance, to letting four million cars go on the market with defective lower control arms by simply saying, 'Well, that's what management wants us to do and that's all there is to it.' Those automotive engineers could have read any number of codes of ethics, such as the National Society of Professional Engineers' canons of ethics. The NSPE code states specifically that an engineer is obliged, after exhausting his internal corporate remedies, to go to outside authorities to alert the public about a hazard in the product on which he was asked to work.

The basic function of a professional society in defending the whistle-blower is to make it unnecessary for an individual to perform an act of courage in order to utter a simple statement of truth. How much courage it takes for one simply to speak one's mind or act on one's conscience is a function of how authoritarian a system is; and the amount of courage needed is an indication of just how authoritarian the process has become. There is no doubt that if we are going to await these expressions of courage not many of them are going to be forthcoming unless the professional societies assume the responsibility for defending their members in such circumstances.

At what point should corporate or government scientists, engineers, or other professionals dissent openly from their employer-organization's policy? If the professional does dissent, how can he protect or defend his decision to place his professional conscience over what he believes is his organization's illegal, hazardous, or unconscionable behavior?

These are important questions rarely answered in the context of controversies such as the defoliation of Vietnam or the standards for constructing nuclear power plants. 'Duty,' said Alfred North Whitehead, 'arises from our potential control over the course of events.' Keeping silent in the face of a professional duty has direct impact on the level of consumer and environmental hazards. This fact has done little to upset the slavish adherence to 'following company orders.'

Employed professionals are among the first to know about industrial dumping of mercury or fluoride sludge into waterways, defectively designed automobiles, undisclosed adverse effects of prescription drugs and pesticides. They are first to grasp the technical means of preventing existing product or pollution hazards. But they are very often the last to speak out, much less to refuse to be recruited to support or perform acts of corporate or governmental negligence or predation.

The twenty-year collusion by the US automobile companies against the development and marketing of exhaust control systems is a tragedy for those engineers, who, minion-like, programmed the technical artifices of the industry's defiance, to say nothing of the cynical and knowing violation of American antitrust laws by lawyers and high corporate officials.

A prime foundation for professionalism is sufficient independence to pursue a course that could save lives, secure rights, or preserve property unjustly imperiled by the employer-organization. The overriding ethic of the professional is to foresee and forestall the risks to which he is privy by his superior access and knowledge, regardless of his own or his employer's vested interests. Physicians should strive first to prevent disease; lawyers should apply the law to prevent auto casualties; economists should try to clarify product and service characteristics in the context of quality competition; engineers should make technology more humane as a condition of its use; and scientists should anticipate the harmful uses of their genius.

All these ideal missions, unfortunately, possess neither the outside career roles for their advancement nor sufficient independence for the organizationally employed professional to exert his conscience in practice beyond that of the employer's dictates. The multiple pressures and sanctions of corporate and government employers are very effective in daunting the application of professional integrity. When, on occasion, such integrity breaks through these restraints, the impact is powerful. This might well explain the organization's determined policy of prior restraint.

During the past half dozen years of disclosures about corporate and governmental injustices, the initiators have largely been laymen or experts outside the exposed system. The list is legion – black lung, brown lung, DDT, mercury contamination, enzymes, phosphates and NTA in detergents, SST hazards, MER 29, and nerve gas storage and disposal. Inside the systems, however, 'mum's the word.'

Three basic changes are needed to make a beginning:

1. Goverments should enact legislation providing safeguards against arbitrary treatment by corporations and goverment agencies of employees who exercise their constitutional rights in a lawful manner. At a minimum, such laws would authorize the courts to protect a professional's 'skill right' in a far more defined manner.
2. Employed professionals should organize to provide a solid constituency for the adoption by management of the requisite due-process procedures which the professional can appeal to or enforce in the courts.
3. As I have already suggested, professional societies should clearly stake out their readiness to defend their colleagues when they are arbitrarily treated for invoking their professional ethics toward the corporate or government activity in which they were involved. Most of the established professional societies or associations never challenge corporate or governmental treatment of lawyers, engineers, scientists, or physicians, as the American Association of University Professors has done, on occasion, for university teachers denied academic freedom. And where there is no willingness to challenge, there is little willingness for the employee to dissent.

To require an act of courage for stating perceived truth is to foster a system of self-censorship and the demise of individual conscience against the organization. Whether a professional's guiding light is a desire to abide by the law, to pursue the Golden Rule, or to adhere to his profession's code of ethics, his paramount allegiance to his fellow man, rather than to an illegal, negligent, or indifferent organization, must find concrete recognition, respect, and defense.

Against Hunger

ALICE MARY HILTON

Against Pollution and Hunger:
Environment and Development

> There are so many hungry people in
> the world that God cannot appear to
> them, except in the form of bread.
> (Gandhi)

A widely accepted (but unproved) assumption holds that human
beings can have *either* a 'good' — i.e. clean — environment *or* technolo-
gical/economic development, but not both. This is connected to a widely
accepted (and also unproved) assumption that all this concern about
'environment' and 'ecology' is just another plot of the rich nations to
keep the poor nations poor.

Both assumptions are tricky because both contain a tiny kernel of
truth, and both are dangerous because, in spite of that kernel of truth,
they are totally false.

To investigate and try to understand the relationship of environment
and development one must, first of all, clarify the terms.

By environment, one generally means an evironment that is good for
people and healthful and pleasant to live in. It is assumed that a desirable environment consists of clean air and clean water. It does indeed.
But around the North Pole the air is clean, the water clear, and the
environment probably fine for polar bears, but hardly suitable for human beings.

A good environment is more than clean air and clear water — though
these are of basic importance. A good environment for human beings
contains good and nourishing food, decent houses, warm and attractive
clothes, good schools and medical services, books and art and music,
and everything that brings beauty and joy into human lives.

Development, we now know, is certainly more than putting ugly
Manchester cotton clothes on 'the natives' and teaching them a few
hymns. It means helping human beings to provide themselves with good
and nourishing food, decent houses, warm and attractive clothes, good
schools and medical services, books and art and music, and everything
that brings beauty and joy into human lives. And certainly that means
clean air and clear water.

27

· So development and a good environment are really the same, and for all our euphemisms they cannot be separated. But neither must we pretend that everything is just as good or bad everywhere as it is everywhere else.

There are differences and there are different problems to be solved in different areas, because we live in a world of poverty and riches, of poor nations and rich nations.

Underdevelopment: poverty and hunger

Let us not mince words: Underdevelopment is poverty — horrible, painful, ugly, heart-rending, body-breaking, soul-destroying poverty. Underdevelopment is children with bellies distended by hunger and legs crippled by malnourishment. Underdevelopment is children dying before they had a chance to live. Underdevelopment is being toothless before twenty and dead before thirty.

What underdevelopment is NOT is romance in the moonlight by an 'unspoiled' lagoon. The unspoiled nature in underdeveloped areas is more likely to be barren rocks, mosquito-infested swamps, impenetrable jungle, or parched desert, all of which can only become lovely and prosperous by development — i.e., the use of sophisticated technology that can bring in topsoil, drain swamps, clear jungles, and irrigate deserts.

No amount of earnest warning by well-meaning and prosperous Western scientists will convince anyone in a poor country that development — i.e. productivity through modern technology — is anything but desirable. And what's more, he is right. It is only the rich who can, and should, indulge in the luxury of worrying about pollution. The poor are concerned with survival. When they become confident that survival is likely — i.e. when their standard of living will reach a reasonably high level, so that at least their food supply is assured — they will surely think about the quality of life. But while there is hunger, there is only life itself — survival — that matters.

Survival and population

It is the concern with survival that has caused concern among the comfortable, for it has lead to what is popularly known as the 'population explosion.' It is debatable whether populations have really 'exploded', but that the number of human beings alive in this world has greatly increased is beyond debate. It is also quite certain that in some areas there is considerable overcrowding — though it is not certain that overcrowding is a result of the increase in the number of people. Overcrowding can be caused by an increase in the population, but it is more likely

28

to be caused by an influx of people from other areas that are getting depopulated. The influx can be due to any number of reasons; it most often is the result of economic pressures, on the one hand, and lures, on the other; and it is the almost inevitable result of the link between employment and income in our society.

Population growth, erroneously confused with overcrowding, is a term that needs to be clarified. Population, as a function of birthrate, is not growing but is failing to diminish adequately. The birthrate — the number of births per thousand of the population — has remained remarkably steady for half a century or more, even in those parts of the world that are of such concern to demographers. In fact, there has even been a marked decrease of births in some areas.[1]

It is a fact, nevertheless, that the number of human beings who populate the earth is increasing geometrically with each generation. Without questioning at this moment whether this is a blessing for the world or a catastrophe, let us investigate why there are more and more people in the world, although it is not true that women have more babies than their mothers or grandmothers did, and often fewer. First of all, individuals live longer, and, in living longer, their lifespan overlaps that of others more frequently — in other words, they are around to be counted more often. This is one aspect of what is called 'death control' that leads to a larger population without any increase in the birthrate. Even more effective is death control at the beginning of life: more babies are born *alive* than ever before in the history of the world and more survive their hazardous first year. Infant mortality is, nevertheless, still far too high among the poor on every continent, and too many children still die before they can reach their sixth birthday.

This is the reason for the geometric increase in the world population — not a 'baby boom', not more pregnancies in terms of per thousands of the population. Children who survive to produce children are, as Cooke reminds us, like untouched savings: 'Each increment of gain is "reinvested" to become part of the "capital".' If the population increases by only three per cent annually and if the divided remains 'unspent', the population will double every twenty-four years.[2] That is simple enough arithmetic — resulting in a geometric increase.

We have becomed accustomed to viewing this increase with alarm, because we equate it with overcrowding and poverty and hunger. We fear, in the future, more hopeless misery and hunger. For what makes the population explosion so very explosive is its concentration in areas that are least capable of supporting even a stable population. Even a slight improvement in the infant mortality rates and the life expectancy rates, even the most rudimentary epidemic control measures, would cause the populations in the slums of India and South America to soar. In Latin

America, for example, the 1920 population of 91 million had grown to 252 million by 1966. At the present rate of increase, it is projected to reach 387 millions by 1980 and 756 millions by the end of this century.[3]

Increasingly, exploding populations crowd into city slums — the *calampas* of Chile, the *villas miserias* of Argentina, the notorious *favelas* of Brazil. In the *favela* of Rio de Janeiro, for example, 400,000 hungry and miserable human beings existed in incredibly overcrowded squalor in 1947. In 1961 there where more than 900,000 – or 38 per cent of the city's total population — in the *favela,* more hungry, more miserable, more crowded than ever, and much more desperate. In Recife, in the North-East of Brazil, half the population is officially classified as slum dwellers. In Chimbote, Peru, three-quarters of the population is in this unenviable category. In Buenaventura, Colombia, it is eighty per cent of the population for whom the 'Good Adventure' is an overcrowded *villa miseria.*[4]

But outside the slums, outside the exploding cities, the vastness of empty continents cries for people. In Latin America it is almost an article of faith that the continent is not yet too densely populated.[5] One need only fly across the United States or Canada or the North of Norway to see the vast expanses of empty land, thousands of miles with scarcely a few houses dotted about. Even on the overpopulated subcontinent of India, where 20 per cent of the world's population inhabits 3 per cent of the world's surface, more than half of the inhabitants are crowded into less than a fifth of the available area. 'In spite of tremendous pressure on certain areas — the valley of the Ganges, the Bengal delta, the Orissa and Cochin regions — a third of India's land resources remain unexploited.'[6]

Why all the overcrowding, the misery? Why the unimaginable hunger that plagues two-thirds of mankind? Humanity is exploring the moon; surely the vast areas of arable land that are uninhabited and unproductive cannot be beyond our ability of exploration as sources for life and sustenance for human beings. The human brain is capable of planning a human colony on a man-made satellite in space; surely the irrigation of deserts is not beyond our imagination. Human ingenuity has learned to make cloth out of coal and strong conctruction material out of glass; surely housing and clothing human beings is not beyond the realm of the possible. At this time of human history, when man has accumulated unprecedented know-how — consisting of vast scientific information and unfathomable technological proficiency — we have, for the first time in human experience, the ability to create unlimited abundance. We lack only the vital ingredients: *the wisdom to do what never could be done in the past and the will to do what we know how to do.*

Education and development

We take for granted that, in our time, education has advanced in great strides, but we rarely define how education has advanced, in what area, in which part of the world, whose education is superior, whether the advance has been a strengthening, a broadening, a deepening, a spread. . . It seems true beyond doubt, that in the rich nations, there is the collective know-how to do many things. But when we look closely, collective know-how is a myth — it is a composite of know-hows that shows great gaps. We know how to grow food in tanks of water and shall probably learn how to grow it on the bottom of the sea or in outer space. But that has not prevented the shocking fact that 'of the sixty million deaths recorded annually in the world, thirty to forty million have to be attributed to malnutrition.'[7]

We know how to build bright and airy homes. But that has not prevented the shocking fact that in Mexico, for example, almost three-quarters of all dwellings consist of one room that is dark and gloomy. Ninety percent of these dwellings, in the rural areas, are without windows or floors, without plumbing or furnishing — 'mud and wattle huts that have not changed since pre-Columbian times'.[8,9] In such squalor human beings are without hope. Yet there they must find shelter; there they prepare their meager food and sleep, there they make love and bear children. In abodes, not much better than the warrens of animals, they exist and die — generation after generation, from its non-existing cradle to its forgotten grave. And only the more fortunate among them will have a chicken or two, or a treasured goat, to share their 'home'.

We know how to make comfortable and beautiful clothes out of 'miracle' fabrics to protect ourselves against sun and wind, against rain and snow, and even against the emptiness of space. But that has not prevented the shocking fact that to the undernourished millions who are so explosively crowded into the world's slums, clean and untorn clothes and shoes are 'miracles' — because they are unattainable.

Can we really speak about great advances in education, when, with all the know-how to create abundance on earth, we have not learned how to assure the most modest decency to two-thirds of mankind? No matter how many children in the rich nations remain in school well past puberty, no matter how many of our young are in our proliferating institutions of prolonged instruction, no matter how many of us pass the required examinations that entitle us to write letters after our names, has our education really reached new heights as long as we can plead ignorance about the majority of mankind, care little about the causes of their plight, and only rarely wonder how the situation can be changed?

And still, we in the rich nations are considered better educated than people in the poor nations. By which we do not necessarily mean that all of us — or many of us or even *any* of us — are more knowledgeable or wiser or better trained than individuals elsewhere. We talk in averages, in terms of school attendance, technical literacy, perhaps, or training suitable for the industrial (but not necessarily the cybercultural) age. And because of our prosperity, we serve poor nations as a model of educational achievement. Without discussing our rights to serve as models, let us look at the chances of the poor nations to reach the educational goals as outlined in one of the most ambitious sections of the Punta del Este Charter adopted when leaders of the Americas met in Uruguay in April 1967. The goals were:

To eliminate adult illiteracy;
To assure each child a minimum of six years education by 1970;
To modernize and expand vocational, technical, secondary, and higher educational and training facilities;
To strengthen capacity for basic and applied research;
To provide the competent personnel required in rapidly growing societies.[10]

The hope to meet these goals was, to say the least, unrealistic. Guatemala, for example, would have to accquire 73 per cent more teachers by 1980 just to have school attendance continue at its present 35 per-cent ratio of the school-age population. If the 65 per cent of school-age children currently excluded from school attendance were to come to school. the number of teachers would have to increase by 358 per cent.[11] Improbable though it seems to achieve such an increase under the best of circumstances (in a little more than one decade), it becomes quite hopeless in a country where dietary deficiencies have reduced most children to apathy and pellagra. How then could each child be assured six years of education by 1970? (Obviously, the goal was *not* reached – or even approached.)

There remains the important question whether six years of schooling is really the most urgent matter in a situation where starvation is the daily, all-absorbing catastrophe in the lives of the majority. What do you teach children who have never seen a glass of milk? How do you get them to school, if they are too weak to walk? Why should they learn to write, when they are too apathetic with hunger to see the world around them? Why learn to add, when the years of your life are likely to be fewer than can be counted on one hand?

We seem to think it is of the utmost importance for every child to acquire technical literacy. But the mere technical skill to read and write and do sums cannot be called education, nor are such skills necessarily the basis of education. Contrary to popular opinion, the rudiments of arithmetic are *not* the first step toward a knowledge of mathematics, but of accounting; and whereas the ability to read and write is of immeasurable importance to carry on and preserve civilization, there have been many cultures that have passed on their history and traditions and their accumulating know-how and wisdom verbally from generation to generation. The effort expended on learning and verbally passing on knowledge of one's tribe is at least as great as the effort to become technically literate, and the chances that some of the accumulated wisdom finds its way into the heads of the young is infinitely greater through personal instruction rather than indifferent training in skills. One thing seems certain: In the evolving cybercultural society it will take more than technical literacy and basic skills to participate in the decision-making processes on any level, and a great deal more education to equip future generations for the Good Life.

But in the age of agriculture, education, including the basic skills, was for the few. Society always needs educated human beings, but agricultural societies need only few. Every agricultural society needs a growing population, for human muscle power was the only source of kinetic energy that was available to produce whatever there was to consume. Ever since the neolithic agricultural revolution, ten thousand years ago, when human ingenuity invented the plough and man turned from a food-gatherer into a food-producer, the need for human muscle power had became the primary need of every agricultural society.

Before the age of agriculture, the food-gathering, hunting tribes could not have survived unless each member possessed skill as well as brawn, brain as well as muscle. But to the agricultural societies that followed, more people — appropriately called 'hands' — meant more machines, and that was equivalent to greater prosperity. Hence, the social ethos in every agricultural society decreed the virtue of large families. Birth control, abortion, homosexuality — anything detrimental to the production of more hands — was condemned. Education for the 'hands', in an agricultural society, is not important. It is, at best, a luxury that the rich farmers can afford, although it was never allowed to interfere with harvesting.* In the vast agricultural holdings of feudal times – and they have lasted much longer in some parts of the world than in others – educated hands are apt to become troublesome.

The industrial revolution presented society with a dilemma. There was no doubt that hands were needed in industrial society. Industrial barons of the nineteenth century were quite as interested in an expand-

---

* A left-over from the agricultural age are long school holidays during the summer, when it was deemed more important to have children help with the harvest than to improve skills in the three R's.

ing population as the feudal barons that preceded them. In the new world they looked to the old as a source for more hands: in the southern United States land-owners bought their hands from Africa; in the northern United States the industrial barons went to Europe. The European industrial barons endeavored to grow their own or to get the worst chores done elsewhere, wherever cheap hands were plentiful.[12]

Unlike farm-hands, factory hands are more valuable when they can read instructions and dials. And the more complex machine systems became, the more urgently needed were factory hands who possessed technical literacy. In the United States, public schools became a necessity, if industrialization was to be successful — the more so because so many hands were imported and had to be taught English as well as the ethical precepts of an industrial, puritan society. So schools were built. For a society that needs literate hands always finds the means to build schools.

In the agricultural societies, technical literacy was not important. Neither the social ethos nor laws required it, nor did the society provide the means to 'educate' the young in the basic skills. But it would be a mistake to believe that the *technically illiterate are necessarily uneducated*. They may be very well educated indeed. But they are usually educated 'only' in the traditions, the history and legend, the music and art of their culture. Taught by their mothers and by the elders of the village perhaps, they have little occasion to learn about other ways of life. Custom is not lightly discarded in such societies, and the social ethos does not change easily where it is learned by patient repetition and committed to memory. In the technologically underdeveloped nations of the world, what has been true since time immemorial is still the social ethos of most populations: a large family means not only prosperity, but status. Even though large families all too often do not mean prosperity any more, they still confer status. And the social ethos still decrees a birthrate as high as it was necessary in the past to assure that there would be an ample supply of hands to till the fields in spite of the staggering infant mortality rate and the unbelievably low life expectancy of those days.

Hunger and population

We may consider it unreasonable of Indian villagers to confer status upon a woman in direct proportion to the number of sons she bears, but the approval of her peers is more important to her than our estimate of what is reasonable. And quite realistically, her own security and that of her husband depend upon having children to support them should they survive beyond the twenty-seven years that are her life expectancy in India to this day.[13] What could be more rational behavior than

34

to have a baby every year, when it assures the mother the approval of her village and her own social security? How can she be expected to violate the precepts of the social ethos she has been taught and to jeopardize her own survival! The preaching by a remote government of vague concepts like 'death control' and 'population explosion' makes little impact. She knows that a quarter of the babies born in the village die before their first birthday and almost as many before they are five years old. She may not know anything about statistics or fathom the disastrous effect on the Indian economy where half the children born consume their starvation diets and die before they can become producing farm hands. But she has seen her own father die when most of her brothers and sisters were much too young to support themselves and when there were not enough older ones to care for their ageing mother.[14]

Economists and governments talk about a lagging productivity gain and a disappointing Gross National Product and hope their people will try to understand that having fewer babies and going to school will still their hunger pains. It must be quite meaningless to the 87 per cent of the Indian population who still live in tiny villages where little has changed in centuries.[15] The land is still farmed, as it always was, with tools that bear more resemblance to the primitive digging stick, invented with such cataclysmic results ten millennia ago, than to the sophisticated machine systems helping to usher in the cybercultural age on other continents. In India, a farmer gets only about a fourth of the amount of rice that an Italian farmer harvests from a field of equal size. The reason is not that Indian soil, only the most fertile of which is farmed, is less fruitful than Italian soil. Not the Indian land but the Indian farmer is exhausted. Hunger and disease have taken their toll and he is no longer an efficient machine. He lacks tools and fertilizer and well-selected seeds. Even manure cannot be fed to the land to help it yield more food, for manure is fuel — the only fuel the poor can get — and it must be used for cooking and warmth.[16]

The ghastly trio — hunger, disease, and exhaustion — is the enemy. But because it is so formidable and frightening an enemy, because we find it difficult to cope with the problems of two-thirds of mankind, we prefer to blame their troubles on their very existence — at least on their existence in such large numbers.

And we wish the problem would go away! But it makes no sense for us, in the rich nations, to wish the birthrate would decline in India and Latin America and China, and every other area of 'surplus' population. First of all, wishing does not make it so. And if wishing could bring about changes in the birthrate that could solve the serious problems of hunger and despair and growing rebellion against their misery by the vast majority of all human beings on earth, then the birthrate would have to

35

be reduced sufficiently drastically to bring about the rapid extinction of the poor, unwanted, exhausted, and obsolete human machines for which modern technology has found more efficient substitutes. If human beings are 'hands' — i.e., machines — then the vast majority is indeed 'surplus', then there is indeed a 'surplus-population' problem. But human beings are *not* just 'hands', they are persons. None is so dull — or so emaciated — that he does not have a spark of the human spirit that some still call his soul. There never was a child born into this world who did not have the potential of being a unique event in the universe. And he could, if his potential were realized, make his unique and personal contribution to human civilization.

The population-growth problem is not a population-growth problem at all. It is a population-hunger problem of staggering proportions. But hunger is not a new problem. It is an age-old problem that mankind has been unable to solve in the past. For no matter how many 'hands' humanity produced, the 'hands' could not produce enough to feed themselves They had neither the strength nor the tools nor the skill to produce enough.

Our technological skill, for the first time in human history, has made it possible to produce abundance. For the first time in history, man's ability to produce food might catch up with his ability to produce 'hands'. But now 'hands' are no longer in demand and human labor is becoming obsolete in a world ruled by the social ethos that laboring is the only means for the vast majority to obtain a license to consume what human ingenuity can produce. Hunger, in the dawning age of cyberculture, is not a problem of demand exceeding the supply. For our new potential to supply whatever we may decide to demand is the true miracle of the age.

Hunger is now a problem of distribution. It is the failure to distribute available resources intelligently and of allocating our potential resources — i.e., the resources we can choose to produce whenever we learn how to distribute them — to would-be consumers. As Kenneth Boulding notes, the civilizations of antiquity 'were endowed with such limited economic surplus that they could not have continued to exist except on a basis of extreme inequality in the distribution of wealth. In the last analysis, all ancient civilizations were only small islands of culture, rising out of an immense sea of poverty and slavery.'[17]

The potential of abundance

Too few hands and too little food — an excess of toil and scarcity — that has been the experience of mankind throughout history. But with the harnessing of natural sources of kinetic energy (the power of the wind and the flow of water, beginning in the Middle Ages, of steam, of electricity, and presently of nuclear energy, in the future possibly of the sun and of the tides) — and with the invention of ingenious tools, power-driven machines, and monitoring and control devices (computing machines), the human condition is being reversed: We now have a surplus of generous proportions. We have the potential to create abundance and the potential to make all the hands we need, not by the conventional biological process but by allowing technology to do the producing. We can have all the slaves — of metal and glass and rubber, or of synthetic substitutes — we wish to make. We no longer need to make slaves of human flesh and blood.

We have the skill and the know-how. But we lack know-why and wisdom. Is the majority of mankind destined to remain hands that are not wanted? Or is there another reason why they should live and share in the abundance created by the human mind?

In the past, there was some excuse for hoarding the small surplus: without it even the tiny 'islands of culture' could not have been created. But with potential abundance for the whole world, it is just foolishness to tolerate existing conditions. For example, the 19 richest nations, with 15 per cent of the world's population and 70 per cent of the world income, consume 85 per cent of the world's raw materials; but the 15 poorest nations with half of the world's population and only ten per cent of the world's income, must exist on the tiny fraction of the world's raw materials.[18] Whatever the justification for such distribution in the past, it does not hold when abundance is possible. It is the characteristic property of abundance that the more equitably it is distributed, the greater the benefit that accrues to all — including the rich nations. For the highly developed, technologically advanced nations of the West suffer from one scarcity: customers for their products. As long as there are poor nations this scarcity will persist. The poor are obviously poor customers. Thus raising the standard of living in the poor nations will not thereby diminish the wealth in the rich nations; it will enhance it. The starvation of two-thirds of humanity is the enemy of every human being on earth; not the high birthrate per se, not even if it were not accompanied — as it still is — by a shockingly high death rate. For whereas in the rich nations the life expectancy at birth now exceeds the biblical three-score-and-ten, in the poor nations it is thirty years. (In India it is 27 years.)

We, in the rich nations, have come to regard birth control — particularly in the Brazilian *favelas,* in Indian villages, and in our own slums — as the panacea for the world's ills and the passport to salvation for the starving. Although it is true that effective death control unaccompanied by birth control must result in enormous increases in the population, there is no conclusive evidence of an optimum population for the world or of any estimable limitation of our productive potential. One thing is certain: the birthrate does not decline with a diminishing standard of living. The experience of humanity has shown the very opposite: namely, that, for a variety of well-documented reasons, the birthrate shows marked decrease as the standard of living is raised. This has been observed under diverse conditions and proved to be true in France and Japan, in Sweden and China.[19]

Moralists in the rich nations have inferred that the poor are improvident, irresponsible, and lecherous. There may be reasons, however, that are less damning for the character of the poor and more biologically and socially conditioned. The correlation between increased protein intake and decreased fertility, for example, is statistically corroborated in the findings of endocrinologists. The Romans had a special word for those who, on a starvation diet, produced *prolific* offspring: the *proles,* or proletarians. In Latin-America, there is a saying: 'The table of the poor is meager, but fertile is the bed of misery.'[20] 'Hunger is responsible for the overproduction of human beings,' says Josué de Castro, 'excessive in number and inferior in quality, who are hurled blindly into the metabolism of the world.' And he shows that where the birthrate is highest — 45.6 in Formosa — the daily consumption of animal protein is lowest — 4.7 grams; but where the birthrate is lowest — 15.0 in Sweden — the protein consumption is highest, namely 62.6 grams per day.[21]

In a series of rat experiments, Slonaker found that when proteins constituted more than 18 per cent of the total calorie intake, sterility increased, the epoch of fertilization of the female was retarded, the number of litters was reduced, and the number of young in each litter decreased. Studying the reproductive indices of groups of rats fed diets with different protein contents for six generations, Slonaker observed that groups on a ten per-cent protein diet showed only five per-cent sterility; when the protein content was increased to 18 and 22 per cent, sterility increased to 22 and 40 per cent, respectively. On a 10 per cent ration, each rat produced an average of 23.3 offspring; on 18 per-cent protein, she produced 17.4, and on 22 per-cent protein, the produced only 13.8 offspring. In other words, by doubling the protein content in their diet, the fertility of rats is almost cut in half.[22] Obviously, the better diet produced stronger offspring and thus reduced the infant mortality rate among the rat population. There is a clear example, observed under laboratory conditions, of relationships in a natural cybernetic system. Feedback is in operation. When nutrition is adequate to ensure high survival rates, the number of offspring produced decreases, but when the quality of nutrition decreases and therewith the survival chances of infants, fertility increases in order to guarantee the survival of the species.

The most cursory studies of human societies show the same conditions, reinforced by social and economic forces which will be discussed later. It would seem obvious, therefore, that if man's salvation lies in the limitation of his expanding numerosity, the most effective means to this end is improving the diet of the hungry majority. That is the most rational. And it is the most humane.

The evidence that populations decrease as their protein intake increases should convince the most spohisticated Westerner, even though he may consider it too simple to believe that the problem of starvation is best solved by feeding the starving. It is true that this is as tautological a statement as the obvious truth that people are hungry because they do not have enough to eat. But it is neither the simplicity nor the tautology that makes us uncomfortable; it is rather the conclusion that, accepting this simple truth, we find ourselves compelled to do something about the situation. The responsibiliy for the shocking conditions 'there' may not be ours — certainly, we did not wish them to arise — but the responsibility for improving these conditions clearly is. Accepting the responsibility for feeding the hungry majority of mankind seems to us to be beyond our capacity. This is where we are wrong. We can eradicate hunger and want in this world, and we, the rich nations, are in the best position to do so. We did not have the capacity to do so in the past. But we have it now. That is a technological fact. We have not yet reached the social maturity to accept it.

The know-nothings and the know-it-alls

In the prosperous, technologically advanced nations, the social reaction to the desperate plight in the poor nations vacillates between the 'know-nothing' and the 'know-it-all' approach. We tend to hide poverty and ugliness. What is behind the well-landscaped green curtains that insulate our suburbs and superhighways, we do not have to see. We have the childish faith that what we can't see does not really exist or, at least, that it can't hurt us. Mindless optimism has been elevated into a virtue and the three monkeys (see-no-evil, hear-no-evil, speak-no-evil) into a symbol of wisdom. We rate the booster highly as an optimist and a patriot and we denounce the social critic as a pessimist and a traitor. (The very opposite is true. The patriot who 'accentuates the positive' is really a man without hope; having no faith in man's ability to succeed in improving the existing situation, he seeks to destroy not human errors but the critic who would expose them in order to help correct them. The critic is the real optimist for he believes that what is wrong *can* be put right.)

At home, the know-nothings seek to ignore poverty and seething discontent behind the green curtains, and feel almost guilty of indiscretion when the evidence is inescapable. On their herded travels abroad they seek the safety of the green oases so thoughtfully — and profitably — provided by Hilton Hotels International and feel a little cheated if the deserts of misery have not been successfully obscured from view. What we euphemistically call 'accentuating the positive', are merely the innumerable collusions of perpetuating the socially-approved lie that all is well in this best of all possible worlds.

Positive thinking is the gigantic lie — the 'heroic' lie — of our society; but many think it is the only possible cure for our insomnia. Particularly in such perilous times, the 'heroic lie is cowardice,' said Romain Rolland, for 'there is only one heroism in the world: to see the world as it is, and to love it.'[23]

The other reaction to poverty and hunger — second in popularity to know-nothingism, but more fashionable, particularly among liberals and pseudointellectuals, is great concern with the misery of the 'underprivileged', particularly those who are thousands of miles away in one of the 'underdeveloped' or 'developing' nations. We are inventive and can create euphemisms at the drop of a hat. And we abhor crude words like 'rich' and 'poor' and 'hungry'. The know-it-alls can see that 'something must be done'. Preferably, something must be done 'over there'. And another conference is called, yet another study group is appointed, another blue-ribbon commission to talk endlessly about the dignity of the poor that would be violated if they were 'given anything for nothing' is set up. To make sure that the poor retain their dignity, they must remain poor until they 'learn to help themselves'.

Like know-nothings, the know-it-alls use language to deceive. They, too, obscure the world because they cannot love it. Their words bear no relation to the stark truth of poverty. For there is no dignity in spindly-legged children dragging their hunger-bloated bellies through the misery of their inhuman existence. 'Helping themselves' is not feasible for men who are immobilized by the apathy of undernourishment. And there is no learning possible for children like the Quechua Indian children who chew 'on cocoa leaves, the source of cocain, to suppress hunger pains'.[24]

The liberals and pseudointellectuals are not indifferent. They do not hide in green-shrouded suburbs or stay in the bars of Hilton International. They like to get things done, particularly 'over there'. At present, they feel convinced — they always do — that they have found two panaceas at once: education and birth control. Obviously, the argument goes, 'they' are illiterate, and the illiterate cannot be prosperous and happy. And, obviously, there are too many of 'them', and, therefore, there is not enough food to go around, and, therefore, 'they' must not remain illiterate and numerous. Therefore: education and contraception. There is the nice and neat solution: education and contraception! The perfect solution! Perfect for us, since we need do no more about it, except perhaps write a rather small check, and can enjoy the warm glow of self-congratulation for having presented the hungry two-thirds majority with such a lovely solution to 'their' problem. The rest is up to them. Obviously.

It is less obvious that the problem is not 'theirs' and cannot be solved by 'them'. The world is an integrated system in which connections are becoming more obvious and more closely knit with every technological advance — and it is the very nature of a system that everything hangs together. Whether we like it or not, there is interdependance of all components in the system. Starvation in India and in the poverty-stricken northeastern regions of Brazil is a world problem that requires a rational world approach. There can be no question of the global moral imperative for active concern about the miseries that plague two-thirds of mankind. It demands action — global action. And those best equipped to act, can and must do most to change shameful conditions. Self-help means just that: there is a global problem and it requires global remedies.

One world

The world does not consist of islands. Our welfare is affected whenever a human being — next door or on another continent — cannot get enough food. There is the moral imperative to assure him enough. There is also common sense and economic self-interest. It diminishes our prosperity when what we produce so abundantly is consumed only scarcely. Abundant production must be balanced with abundant consumption. Where the latter falls short of the former, the result is wheat rotting in vast granaries — at the expense of the rich nations who must pay for the folly of allowing the poor nations to starve. The result of such imbalance is not only the expensive waste of artificially preventing abundance from being produced. We pay farmers for keeping fertile land fallow, efficient machinery is rusting in barns; factories are kept from producing up to their capacity 'because there is no market' for goods that millions of human beings clamor for. For the starving in the poor nations — and in our own poverty-sullen slums — survival depends upon our effort to produce to our utmost capacity and to invent the most practical and efficient means of distributing what

41

we can produce. But on the very same effort depend also our own prosperity and wellbeing and the peace of the world.

A world that can be circled in a few hours, where, via television and radio and telephone, everyone's way of life is known to everyone else, is too small to be divided into the rich nations and the poor nations. Human beings who live within talking and seeing distance of one another are too close to be separated into those who cannot eat and those who cannot sleep. The cure for prosperous insomnia is inextricably bound to the cure for starving inertia. For only a world in which every human being has enough to eat is a world in which every human being has a chance to sleep.

World trade and world exploitation

Let us look at 'the march of events' and what it makes clear. The gap between the rich nations and the poor nations is, without a doubt, widening. Gunnar Myrdal compares the long-term development of prices of primary and manufactured goods and finds the former increasingly lagging behind the latter. Since the export of primary goods – including much-needed food – accounts for 90 per cent of the foreign exchange earned by underdeveloped countries, the vicious downward spiral gains momentum. *The poorest must export more and more for less and less.*[25,26] According to a United Nations study of the forty-year period between the end of the 19th century and the outbreak of World War II, the same amount of exports from the underdeveloped countries at the end of this period would buy only 60 per cent of the imports it would have bought at the beginning.[27] It is perfectly logical, therefore, that 'the underdeveloped countries helped to maintain . . . a rising standard of living in the industrialized countries without receiving . . . a corresponding equivalent contribution towards their own standard of living.'[28]

With continuing and increasing disparity, trade between the rich nations and the poor nations would have come to an end long ago, if it were not for credits and grants-in-'aid'. The situation is somewhat like a poker game in which one lucky player pockets all the chips. When everybody else is broke, the winner must either let them have some chips, or stop playing. But under such conditions, the player in control of the chips is in control of the game. It is then preposterous to speak of a game where everybody takes an equal chance on the cards and 'let the chips fall where they may.' Nor can we speak of free international trade when credit and aid are controlled by the rich nations and the poor nations are forced to sell more and more for less and less, while their own needs increase and their resources dwindle. For decades the problem has been debated. In December 1952, the General Assembly of the United Nations adopted a resolution – one in a series of resolutions adopted before and since – regarding the 'financing of economic development through the establishment of fair and equitable international prices for primary commodities.'[29] Resolutions have no effect unless they are acted upon. If any of these resolutions had been acted upon, their impact upon the hungry of the world would have been enormous. In the study of the United Nations Economic Affairs Department, it is estimated that a ten per-cent improvement in the terms of trade in favor of the underdeveloped countries, 'would modify their capacity to import by as much as $1.5 billion a year'.[30] A staggering amount for the poor nations, but no more than the United States is spending upon the slaughter in Vietnam in two weeks.

Such an improvement in the terms of trade might also mean that the poorest nations might not have to export quite so much of the commodities they need so desperately to feed their own starving people, or that some of the land now devoted to exportable crops might be used to grow a variety of foods for native use. To accomplish this, a regulation of trade terms would, of course, have to be accompanied by measures that would assure such results. There is little chance to improve the lot of the hungry, if trade terms are modified only to increase the already enormous incomes of the few owners of the vast latifundia. Like birth control and education, trade agreements and regulations are not going to serve as the panacea. The problem is complex – the solution will have to match it in complexity. Before we can attack the problem, we, in the rich nations, will have to understand how vast the chasm is that separates the poor from our affluent world – and how far they have to go before they reach a standard of living comparable to ours.

We have already shown the unequal distribution of the world's people, incomes and resources. Such a distribution pattern has predictable consequences that seem obvious or surprising to the conventional wisdom. Overcrowding, for example, is not necessarily due merely to the large populations in the poor nations, since some of the poorest nations, for example India and Brazil, have vast territories that are virtually uninhabited. In fact, poverty used to be considered the direct result of *under*population. Crowding, now, is directly attributable to a lack of resources to develop resources. This may take many forms. In Central America, for example, the land is exploited without any responsible policy of conserving its fruitfulness: it yields rich banana crops for seven years and is abandoned. Some areas need water; some must be cleared. Some have never supported human beings; some were the breadbaskets of the world centuries ago, but have not been fruitful, for a variety of reasons, since. Most tragic is the plight of human beings who, sometimes for generations, have managed to survive on land that has yielded too little to nourish them adequately. With modern agricultural tools and methods, it might become fruitful. But generations of undernourishment and malnutrition have taken their toll; in the end, exhausted people rather than exhausted land has been the cause for deserting the land and increasing the crowding in city slums where existence is horrible beyond words but survival has still a greater probability than in the parched, neglected, empty lands.

Continuous movement into the city slums makes the shockingly poor nutrition among the poor in the poor nations even worse. One-crop economies — of crops grown for export — that the poor nations consider necessary, often for the best motives, for example in order to import capital equipment, further contribute to poor nutrition. The poor who work on the coffee plantations of Columbia or the tobacco plantations in our US southern states invariably suffer from malnutrition because their diets are incredibly monotonous and lacking in vitamins, minerals, and animal proteins. Instead of cultivating their own gardens and improving their diets by raising a few chickens or keeping a goat, land and people are devoted to producing exportable crops. People are fed whatever is most cheaply available — usually corn, or rice, or some other carbohydrates, but rarely vegetables that would take space from the tobacco or coffee crop, and almost never meat or fish, milk or eggs, which take space and time and energy which can't be spared in the struggle for foreign exchange or the profits of absentee owners of the vast *latifundia*. Well-meant or not, efforts to increase the well-being of people and their nutrition by ill-considered industrialization have often increased only malnutrition. Here a lack of understanding the meaning of cybernation is as much to be blamed as any other motivation, even greed. As an example, let us take a vast chemical plant that was recently built in the poverty-stricken northeast of Brazil. It was supposed to serve a multiplicity of purposes: bring foreign exchange, of course, aid the sugar plantations, and — most important, so it was thought by well-meaning economists who preach the gospel of industrialization without ever investigating modern production methods — *to provide employment.*

The full-employment mania, which cannot be discussed in any detail in this essay, is spread about the world, accepted on faith and stubbornly clung to in spite of countless disastrous experiences.[31,32] In this case, employment was to be provided both directly in the chemical plant and indirectly by the boost in sugar production which could be used by the chemical plant. In fact, no employment followed. With an assured market for their produce, the sugar planters found it expedient

44

to invest in modern farm machinery, which resulted in less, rather than more, employment on the plantation as well as in the refineries, where modern machinery was also installed. And anyone who has seen a modern chemical plant will know that it employs absolutely nobody, except a few top-level engineers, chemists, an administrator or two, and salesmen. Everything is done by machines or, to be precise, by a cybernated machine system, i.e., machines monitored and controlled by a computing machine. Such a machine system in a modern chemical plant does not, and cannot, co-exist with human 'hands'. In this particular case, in Brazil, the effect of the new chemical plant on the lives of the poor was not only no improvement of their lot, but sheer disaster. The people of the area had subsisted on corn and fish. Barely enough corn and an inadequate supply of fish (that was not very good, but did supply some protein to their diet) had kept the population alive. Chemical plants find it cheapest to dispose of their considerable waste products in nearby rivers. (Actually, these 'waste' products are rather valuable; but it would require substantial investments to salvage them. Thus most industries, considering *this* year's — rather than long-range — profits, waste and pollute.) Rivers for dumping waste products were among the chief reasons, together with the vicinity of sugar plantations, for building this chemical plant in this particular area in the remote northeast of Brazil. For the rivers in less remote regions are already too polluted, and the more sophisticated people in the industrial nations with the *most* polluted rivers are already too incensed. The river, from which the unfortunate people had got their meager protein rations, was polluted in no time, and the fish died. And the people, few of whom could get to the far-off *favela* of Rio de Janeiro, died of hunger.

It is not surprising that only 17 per cent of the world population obtain what is regarded as an 'adequate' protein ration, namely one ounce of animal protein per person per day.

For 21 per cent of the world population this ration is reduced to something between one half and one ounce per day. But the majority of human beings – 58 per cent of the world population – have less than half an ounce of animal protein per day.[33] They almost never eat meat, and fish only rarely; nor do they have milk, cheese, or fresh eggs. Half an ounce of animal protein per day would amount to a very, very small hamburger once a week — and no other proteins, such as milk, cheese, meat, or fish at any other time.

We, in the rich nations, cannot really grasp what concerns the poor. For, as Harold Wilson writes, 'For the greater part of mankind, the most urgent problem is not that of war, nor that of communism, nor that of the cost of living, nor that of taxes: *it is the problem of hunger.*

45

And that is because hunger is at the same time the effect and the cause of the poverty and the misery in which one and a half billion human beings vegetate.'[34]

Our failure to bridge the gap between the hungry and the sleepless is the greatest danger to world peace and universal abundance. And basically this failure is caused by ignorance, but particularly this failure must be blamed upon *the ignorance in the rich nations,* where we cling to the comfort of the conventional wisdom — whatever the cost — although we, unlike the poor, do have the means to learn. We, in the rich nations, may have ample reasons for our insomnia, but that is rarely what keeps us awake. We worry, instead, about slogans and imaginary dangers that are part of the conventional wisdom, although most of them never existed but in our insomniac nightmare. The real dangers — most of which *can* be averted — concern very few of us. Take for example the spectre of communism that we seek to exorcise by absolutely fantastic expenditures of resources, the most tragic of which is thousands of human lives, although no idea has ever, in the history of humanity, been defeated by guns. The desperate cannot be frightened, and freedom and democracy have a hollow ring to those with empty bellies. 'Hunger', says Josué de Castro, 'was the great recruiting officer of the army of Mao Tse-tung'.[35] History has confirmed him on every continent.

Making the deserts bloom

Men, who can make a wasteland of a fertile tropical paradise, can also make the deserts bloom. The breadbasket of the Romans has been devastated into a desert, but the arid waste between the Sierra Mountain ranges and the Pacific Ocean has become a garden.

Nowhere has change been more dramatic than in the land where three of the world's great religions were conceived. 'Israel', writes Josué de Castro, 'has regained in the twentieth century the flourishing aspect of the biblical land flowing with milk and honey. The Negev Desert, which covers more than 60 per cent of Israel's territory, has become a splendid oasis.[36] Like Moses, millennia ago, modern science together with courage, intelligence, education, and hard work, coaxed water out of the dry heart of the desert. Against the lack of raw materials, poverty of soil, and unfavorable climate, man pitted his creative imagination. And he won.

On only 7722 square miles of land — inhospitable and wild only two decades ago, when it was inhabited by six hundred-thousand hungry human beings whom it could scarcely support — more than two million human beings are now very adequately fed and comfortably housed. Here

46

civilization flourishes, music is made, the arts thrive, and learning is esteemed — luxuries of which the half-starved people of the parched northeast of Brazil cannot even dream. For the luxury of creating civilization requires the production of surplus. And in Israel there is still surplus food for export. Here the miracle was wrought. Here man made everything out of nothing. Here Vogt's theory of the 'biotic' potential of soils[37] was disproved. Here the myth that man cannot survive in the desert was exposed as a fraud. Here the march of events is standing fast against the conventional wisdom of the neo-Malthusians. Here is the evidence that 'overpopulation', as an absolute value, is meaningless. Three starving people in a desert are overpopulation. But this same desert through the catalyst of human ingenuity — and energy and irrigation — may be *under*populated with three thousand — or three million — prosperous human beings.

It would be naive to think that human ingenuity unaided by material investments can turn a desert into a land of milk and honey. But it would be equally naive to think that material investments, of cash or capital equipment, alone can accomplish this miracle. Israel benefited by the influx of talented and intelligent human beings simultaneously with the influx of capital. There can be no doubt of that. But Israel benefited because of the conditions of and the purpose for this confluence. Human talent came to Israel not to exploit and leave nor to exploit and stay, but to make a home for human beings by its own efforts, much as the European settlers came to the North American continent three centuries ago. And, most important, capital was invested in Israel not, as is usual, for the optimum financial return, but for the extraordinary purpose of providing human lives for human beings — without thought of financial gain. Cynics have said it was conscience money or insurance to keep the Jewish refugees 'over there'! Perhaps. Economists have charged that every orange produced in the Negev Desert is worth its weight in gold. Perhaps. The point is not the precise mixture of noble and ignoble motives that makes human beings commit a generous act. Nor is efficiency equivalent to dividends in cold cash earned by an investment. The success of this investment is in the creative work and the joyful lives human beings can lead and in the happy laughter of healthy children who have good food to eat. If that is foolish sentiment, so be it. But if it is pragmatism to invest surplus — and all capital is surplus — merely to beget more surplus to be cautiously invested where it will beget more surplus (and the conventional wisdom of economists insists it is), then we shall not have a world fit for human beings until we permit foolish sentiment to work its miracles and bury such sterile pragmatism forever, before its cold and bony hand squeezes all life and joy from this world.

Economics and abundance

The miserly hoarding of capital for the begetting of capital is nothing but the atavistic remnant of the age of toil and scarcity. We have come to the threshold of a new age, an age so bright with promise that the weak-eyed miser must hold up his hands to shade his myopic eyes. His conventional wisdom is to hoard and hide and save for a tomorrow that might be poorer; he makes do with as little as possible. We must not judge him harshly, for in the centuries when the scarce morsels that man could produce was all man had to consume, the capitalist miser served humanity as well as anyone could. He hoarded capital and turned his back on misery, but he did increase enormously the productivity of the earth, and he challenged the ingenuity of human beings. He left misery unconsoled; he may have contributed to much suffering; but, whether there might have been better pilots to guide mankind through the sea of scarcity or not, he brought us to his new era that he himself cannot understand.

The new era is the age of abundance: for the first time in human experience man can produce whatever he wishes to consume. Always before our ability to consume has been determined by our ability to produce. *Now* our ability to produce is limited only by our ingenuity in consuming. That is a staggering discovery in a world that has scarcely had time to become accustomed to man's status as a food-producer. We have barely begun to irrigate the deserts and make them into gardens. For every Polynesian island, there are yet a hundred jungles to be pruned — not wantonly razed. For every Negev Desert, there are a hundred wildernesses to be tamed. Man has not yet attempted to plant the icy wastes of Antarctica. He is at best a hunter in the oceans, not a farmer. He has scarcely begun to explore his celestial neighbours. And he has barely thought of conquering the atavistic terrors of his own mind and the fears of his troubled soul.

And yet, it is clear beyond any doubt that man can coax abundance from an eager world. He need never fear scarcity again. But how to conquer his fears is still man's greatest challenge which we call — to hide our embarrassment — the problem of distribution, of capitalization, and of providing incentives and rewards. And then, of course, there is the strong atavistic fear that man is born to suffer, that this world must always remain a vale of tears, and that abundance is sinful and, therefore, must be punished by pollution and death. Against such somber fears, often cloaked in 'scientific reasoning', it is hard to argue. Those who make a religion of suffering refuse joy and hope.

That grave problems exist cannot be doubted. Primarily the problems are problems of distribution, or of economics, of political ineptitude, and of fear of change. Gunnar Myrdal writes:

> A study of trends and problems in the fields of international economic integration ten years after the end of the second World War, must invest us with humility and even anxiety ... The practical problems facing us, if we want to change these trends, are momentous. ... I know of no government and no political party in any country which is really facing up to these problems. There are small groups ... have acquired the strength to see beyond the more immediate things. And there are in all parts of the world many individuals in the great humanistic tradition who also see what the inevitable conclusion must be. When such persons accept political responsibility — and some of them have responsibility of the highest order — a condition for continuance in power is that they accept and recognize the practical possibilities at hand in their several countries, determined by their fellow citizens' present attitudes, and so reserve their profounder insights for general pronouncements that do not upset practical affairs. ... I get the feeling — and sense of anxiety — that the world is drifting towards a destiny that it has not charted in advance and for which it has not been deliberately steering its course.[38]

This is what Lord John Boyd-Orr sees as the *White Man's Dilemma,* as he entitled his book, when he writes: 'He can attempt by force to maintain military and economic supremacy, in which case he will be involved in an almost world-wide disastrous war ... the final outcome of which will be the downfall of Western civilization. On the other hand, he can ... use his present industrial supremacy to develop the resources of the earth to put an end to hunger and poverty with resulting world-wide economic prosperity — in which case he would lose his superior power.[39]

It is this consequence of the second alternative that frightens people in the rich nations, because in the conventional wisdom losing one's superiority and economic supremacy is equivalent to becoming inferior and poor. The conventional wisdom sees the world as a seesaw: if you can't stay up, you must inevitably fall down. What a simplistic view of so complex a cybernetic system! The world of abundance isn't as simply two-dimensional as the dynamics of a seesaw. Complex and intricately interwound relationships do not move in the simplistic up-down pattern of one-to-one cause-and-effect correlation. The world in the Age of Cyberculture moves in an elegant spiral upwards or down; the entire system will move toward abundance or the abyss.[40]

The moment of truth is at hand, for, Boyd-Orr writes:

The Western powers are faced with the rising waves of revolt in Asia, Africa, and Latin America against poverty. They can try to resist it by force or keep it off by the offer of technical assistance and trifling loans with political strings attached to them that will break at the first strain. In this case they will either be destroyed or submerged. On the other hand, . . . they could recognize the inevitable and use their overwhelming industrial superiority to create a new world of plenty. In so doing they would gain new power and prestige by assuming leadership in the march of the human family to the new age of peace and prosperity and the common brotherhood of man, which modern science has made the only alternative to the decline and the fall of the Western civilization.[41]

In the decade and a half since this – among many similar – warnings was sounded and the only possible alternative voiced, there has been little wisdom displayed by the rich nations. Although they have enlarged exponentially their potential for realizing more rapidly than anyone could have predicted 'a new world of plenty', events have repeatedly proved the tendency to 'try to resist by force or keep it off by the offer of technical assistance and trifling loans'. And the former has occurred more shockingly than the latter. For in the last fifteen years, force has been used in Indochina and the Dominican Republic, in Watts and Chicago, in Czechoslovakia and Hungary, and threatened in Cuba, to cite only the most spectacular examples. The US and the USSR are not the only culprits, but they are the biggest and the most dangerous. And they have committed the most monstrous blunders. Against this, how trivial has been the giants' assistance: the Peace Corps, the Alliance for Progress, the Aswan Dam.

Is there still time to 'recognize the inevitable' and to regain the prestige lost so tragically – and deservedly – in the last decade? We cannot be sure. Certainly, it will be more difficult now that the world has lost its trust in the big powers and the gap between rich and poor has widened. Without doubt, it will become more difficult tomorrow. And no one knows when it will be too late, when we shall have crossed the Rubicon or whether, even now, it is already behind us. When Lord Boyd-Orr warned us that trifles would not avert disaster, a jeep could be bought for the value of fourteen bags of coffee; but with the price index in the rich countries that sell jeeps and trucks and machines rising, and the commodities — coffee, bananas, peanuts — that the poor countries must sell, rather than feed to their own peoples, remaining at the same price or even declining, three times as many bags of coffee are now required to buy the same kind of jeep. 'It has been calculated,' states Josué de Castro, 'that the profit thus realized by the rich countries that import these commodities [and export the manufactured products] is equal or superior to the $8 billion paid out in aid by them to the Third World. Aid, then, is . . . simply poured into a pit whose bottom is being dug deeper at the same time!'[42]

In 1964, Lord Boyd-Orr, then Director General of the United Nations Food and Agricultural Organization, submitted to the governments of the United Nations a proposal to create a World Food Board 'to provide financial and other technical arrangements necessary to convert human need into effective demand in the markets of the world'.[43] The proposal was never approved and the FAO for lack of specific powers has never been able to realize any of its creative ideas. It has been no more than a consulting agency in a world crying out for action, but in its files are numerous plans that both rich and poor nations would find of mutual economic interest.

'It is quite possible to make the transition from a colonial economy to a co-operative world economy based on mutual interest without the imperialist or colonizing countries going bankrupt,' says de Castro.[44] The US emancipation from the colonial state has set one precedent; for when the thirteen colonies became the United States they became a better market for British products than before. 'A growing US,' says Earl Parker Hansen, 'did as much as India to create Britain's Victorian greatness.'[45]

Barbara Ward warns that 'some general strategy is needed' for 'patchy development, a little here, a little there, does not lead to sustained growth'. She suggests that the rich nations should allocate one per cent of their Gross National Product to developing the poor nations, for 'the scale of aid must be adequate'.[46]

The cost of abundance

There can be no doubt about the absolute imperative to feed the hungry, to clothe the naked, to house the homeless, to treat the sick, to educate the ignorant . . . to make life worth living for all human beings on earth.

The question uppermost in the minds of all concerned is the cost — not the cash price, but the cost. What about the depletion of our natural resources? What about pollution?

It is held to be almost axiomatic that the cost of prosperity is pollution. The conservationists believe it with horror and sadness, the 'progressives' with a sneer. Above the Arctic Circle, in Bodø (Norway), I found to my astonishment that when certain winds prevail, certain

51

smells become pungent. I was told about industrial plants across the bay and that what offended my nostrils was a fine thing: it was the smell of money. I also heard the sound of jets roaring through the still and bright arctic night, and I was told that what I heard was the song of prosperity. A few students protest, but students, I was told, do not understand the realities of life and the prosperity that comes with factories and NATO bases. The worst is that this is true and that factories and NATO bases do spell prosperity, and that it is better to have a nice house and good food and a brand-new 300-bed hospital with a helicopter port on its roof than to be poor — even at the cost of bad smells and noises. 'One soon gets used to that,' I was told.

In the poor countries, people and their leaders scoff at our concern about pollution. In Havana, I was proudly shown the smoke poured from factory chimneys polluting the clear tropical sky. In India, any mention of the detrimental effects of DDT is met with the angry retort that DDT has saved the lives of 4 million human beings who would have died of malaria, if DDT had not been used.

These are legitimate, serious, and important answers to our legitimate, serious, and important concerns. The question is: Is it true that the cost of development, i.e., prosperity through industrialization, is pollution? Must we choose between starvation for many and an earth that will soon cease to be fit for any living beings? Or can our earth support its human children in comfort and abundance and still remain green and lovely and a joy forever?

The question is: Must the majority of mankind live, as it always has, lives of quiet desperation? Or can human intelligence, which can find a way to the moon, also find a way to feed the hungry, to clothe the naked, and to house the homeless — and do it without destroying the earth?

What about depleting natural resources? What about pollution of the water and the air? What about overcrowding and overpopulation?

There is little evidence to support the supposition that we must deplete our natural resources if we are to have abundance for all mankind. Whenever we deplete, we do so because we waste or because we rely on exhaustible rather than inexhaustible resources. There is no reason or excuse for using up one acre of woodland for every edition of the Sunday newspaper. We have a right to ask whether the Sunday newspaper is worth an acre of woodland. We also can ask science and technology to go to work and find us some way to make paper without depleting the woods. We also have the right to expect legislatures to insist upon protecting our woodlands by limiting waste and insisting upon replanting.Woodland is an inexhaustible resource, *if* we so wish it.

The most important factor in development is, as it always was, energy. All kinds of energy. The question is again whether sources of energy are exhaustible and whether the production of energy *must* cause pollution.

First of all, human intellectual energy is required. This is obviously an inexhaustible resource, but one we have had little control over. Whether the ratio of talent in a population is fairly stable or can be increased has never been established. But it does seem reasonable to assume that the better a baby is fed the better his chances that he will grow into an intelligent person. Enough studies have been made of the effect of early nutrition on intelligence to warrant the assumption that mankind might improve with prosperity.

Second in importance only to human intellectual energy is kinetic energy, or power. Statements made by some self-styled pollution 'experts' that 'all power pollutes',[47] are simply silly.[48] What is polluted by putting up sails to catch the wind and thus channel the energy of the elements for man's benefit and purpose?

To avoid the depletion of our natural resources, it is necessary to phase out the use of fossil fuels to produce energy. These are prime pollutants. Instead we must explore the feasibility of using inexhaustible resources. Of course, I do not mean that we should return to a windmill economy I leave that to the romantic return-to-nature addicts. Everybody else understands that a windmill economy could not produce enough to feed a hundreth of the people who now populate the earth.

It does make sense to seek the means of controlling and using the power of other virtually inexhaustible resources. The power of the tides, for example, is as inexhaustible a resource as the moon. Tidal energy is being successfully used in a powerplant in Normandy. Solar energy will be available as long as the sun lasts, which it is likely to be for a considerable period in the future. If we were to spend a tiny fraction of the financial and intellectual resources we spend so generously on space research of a less promising sort, we might have all the cheap and non-polluting power we are ever likely to need. These are not all the possible, inexhaustible, and non-polluting sources of energy; I have not even mentioned nuclear fusion, which seems no longer fashionable, but cannot be assumed to have failed to live up to the promises made by physicists a few short years ago. Why is so much spent on nuclear reactors that seem so full of danger, and so little on research of fusion? Has fission already become a vested interest — as coal was yesterday, and oil is today?

It is obvious that the allocations of research funds are made for economic-political reasons rather than technical promise. Enormous amounts of capital are invested in oil. Vast financial resources and (political) power interests are at work to keep oil king of the earth. Money and (political) power are great forces in our world. But they are the forces of yesterday because they belong to the age of scarcity. They can be deposed, *if* sensible human beings are willing to investigate the truth and refuse to be sidetracked by light and transient causes. The few who wield power are not nearly so dangerous to the welfare of mankind as the many who wield the multiple banners of reform. The reformers go happily off at a vast variety of tangents, while those who really wield power in this world keep to their well-chosen and tenaciously pursued path.

The trouble in this world is that those who mean well are so often blinded with pity for the dreadful suffering of humanity. They pass out bandaids to deal with a cosmic hemorrhage. And it won't work. Nor will binding little scratches do any good as long as the severed artery is ignored. Neither the miserly conservation of fossil fuels, nor the fearful bowing to an imagined necessity to choose between want for the millions (usually those 'over there'), on the one hand, and pollution, on the other, can make this world fit for human beings. The open-minded search for new (and feasible) ways to produce abundance and for inexhaustible, non-polluting sources of energy can, indeed, make this a world in which all human beings can learn to live human lives.

The distribution of abundance

'The archenemy of us all is the financial system,' writes Colin Hurry.[49] The financial system of capital accumulation for the sole purpose of investment in ventures that yield a profit, i.e. more capital, that can be invested in further profit-producing ventures is the very root of our difficulties. It is an old system, of course, and only rarely questioned because its worth is held to have been proved by time. It is a system that has served civilization reasonably well for thirty centuries, that is true, but they were centuries of scarcity, when some surplus for profitable reinvestments had to be wrung from a very meager world. Besides, there never has been a time like ours when investment merely for the sake of profit was the primary motive for all commerce and manufacture. There was a sense of adventure, in the past, that drove the Fuggers and the Medicis. There was a sense of beauty, and a sense of pride in excellence. But the daring entrepreneur has been supplanted by God 'Efficiency' in the form of management committees. And if there are qualities management committees must never, never display, these qualities are a sense of adventure, a sense of beauty, and a sense of pride in excellence. Management committees of today's super-corporations are not free and daring entrepreneurs who risk their fortunes on the sturdiness of a ship and the excellence of its captain. A management committee first of all risks other people's wealth — if not fortunes, then 'investments'. But of course they do risk their own fortunes too: their promotions, their jobs, or, at least, their status. But, most important, a management committee is not free to do anything, because it is completely and totally dependent upon the financial power structure that wields absolute power behind the business and political thrones of the world. The financial system, the big bankers, rule the world, and all the political and management bigwigs are puppets.

Money is king. And what a grizzly monarch he has become. Unable to comprehend the spring of abundance that wants to burst loose in this world, he keeps his gnarled fingers on the dike of plenty that is about to burst. Money is a merciless, yet a poor old, king, who is afraid of this marvelous revolution of abundance that would — and must — depose him.

King Money is the cause of pollution. For were it not for the required investment, many an industrial plant would gladly cease polluting. There is valuable stuff being wasted in the air. But the management committee dare not spend the money — and thus reduce today's profits — for tomorrow's greater gains. And who can blame them, while money is King! They would not be in management tomorrow, if they allowed today's profits to decline!

King Money is the cause of underdevelopment.[50] For were it not for the required investment, many a poor country could be productive within a few years. Could there be anything poorer than the Negev Desert of yesteryear? Energy and effort made it bloom. The one example of money being expended, not for further profit, for further investment, but for building a good land, has been magnificently successful.

And yet, we do not learn the lesson that is begging to be learned. The financial system is obsolete and intolerable. Investment for profits that are used only for further investments for profits is dangerously foolish. Obviously, shoddy products sold for high prices bring better profits than quality products sold at low prices. Obviously, if the profits from shoddy products are invested to produce more shoddy products for more profits to be invested in the production of more shoddy ... There is an endless and vicious cycle. It causes waste and pollution. It uses irreplaceable raw materials. It consumes energy produced (profitably!) by fossil fuels. It lowers public taste. It enhances consumerdom and destroys creativity. In short, it is the inevitable downward spiral that must lead to perdition.

But the cause of perdition is *not* the population explosion and it is *not* the reasonable and legitimate desire of human beings to live in comfort and joy. The cause of our current catastrophes is an obsolete financial system that is about to choke the very breath of life from this earth.

I must make clear what ought to be obvious, but, astonishingly, all too often is misunderstood. The financial (money) system is not limited to the capitalist nations. If it were, Cuba would not have to introduce the terror of labor camps to keep its sugar production as high as possible. Cuba does not produce sugar for Cubans, but for export. Cubans ought to have more vegetables and rice and fruit. But the land is planted with sugarcane for sugarcane is a cash (money) crop and Cuba needs money to buy tractors, to grow more sugarcane, to earn the money to buy more tractors, to grow more sugarcane, to earn . . .

The same senseless, stifling spiral! The same cause! The outworn, obsolete, non-functioning financial system. This is the archenemy of us all. Let us expand our energy and our intelligence in finding a way to retire King Money and put a worthy heir on his throne. But as long as King Money reigns, let us have a look at him.

There was a time when money was gold. Money, we are told, must function as a measure of value, a medium of exchange, and a store of value. The last, of course, is nonsense, for wealth is a flow, not a store. A flow, by definition, cannot be stored. How can gold be a store of value? Take the money out of gold, and where is its value? Not an ounce of gold would be mined — particularly at the price of gold that does not pay mining except by underpaid, underprivileged South African blacks. Anyway, there is enough mined gold in the cavernous vaults of the world to fill the teeth of untold generations and gild their teaspoons for centuries.[51]

Paper — currency and, even more important, bank checks and creditcards — has taken the place of gold as a medium of exchange. It is more convenient (particularly for tax collectors) and serves just as well. In fact, were its issue properly directed to the end of maintaining fair prices and adequate incomes, paper would serve admirably as a medium of exchange until abundance makes all book-keeping as superfluous as measuring the air one breathes. (And if things go on as they have, we might yet measure and tax the very air we breathe. It is really not much more foolish than exacting payment for the food we eat, which is as necessary for the maintenance of life as air).

But money, paper currency, is not issued by the representatives of the public, who are charged with the responsibility for looking after its welfare. Money is issued by private banks, like the Federal Reserve Bank in the United States and the Bank of England (nationalized in 1946, but not greatly changed in its effect), in the sole interest not of the community but of the *creditors* of the community.

No wonder the miracles of science and modern technology have been of limited blessing to mankind. Civilization has been, in its most vital interests, not in the hands of those who have contributed most to its wealth, but of those to whom in a very literal sense it is indebted and is most likely, if the system continues, to become ever more indebted. The interesting question is: What happens when the Great Creditor has got hold of all the chips? Does the game grind to a screeching halt? Or does the Great Creditor grudgingly lend a few chips to selected players, i.e., those most likely to lose to him and conscientiously and uncomplainingly pay him the usurious rate of interest?

That's precisely where we are now. And we are in this spot because we think there is no other game but the one played with the Great Creditor's chips.

The job of sensible human beings — and I hope scientists are sensible human beings, for their training has been in clear and honest observation and, hopefully, in logical deduction — is to call off the game. Or to get the millions of weak and sheep-like players to see that no natural laws exist to force them to continue to be fleeced.

A distribution system should work properly by distributing what intelligence and energy can create. What about incentives and rewards? They are not so important in an age of abundance when there is plenty for everyone. But if we must worry about incentives and rewards, I have a suggestion: The Marxists used to say (do they still?): 'From everyone according to his ability; to everyone according to his need.' All well and good, and probably too noble for mere men. But in the age of abundance, when machines toil and human beings live, it is quite realistic, I believe, to order the world so that the distribution system provides: *To everyone according to his ability* (to consume — some people really are better at using things sensibly than others); *from everyone according to his need* (to contribute). For some human beings seem to be made to lie under a banyan tree and dream, and others must spend their energy in helping to make the world's wheels go round. And this will be a world of joy and beauty, when the dreamers are allowed to dream and the doers to do — each according to his need

CITED REFERENCES

1. Robert C. Cooke, 'Punta Del Este, 1961–1967: Early Dawn of a Demographic Awakening,' *Population Bulletin 23*, No. 3, June 1967, p. 79.
2. *Ibid.,* p. 55.
3. *Ibid.,* p. 54.
4. *Ibid.,* p. 69.
5. *Ibid.,* p. 58.
6. Josué de Castro, *The Geography of Hunger,* Boston, Little, Brown 1952, pp. 172—174, 179.
7. Josué de Castro, *The Black Book of Hunger,* New York, Funk and Wagnals, 1967, p. 13.
8. Robert C. Cooke, *op.cit.,* p. 70.
9. Inter-American Development Bank, Washington D.C., Social Progress Trust Fund, Fifth and Sixth Annual Reports, 1965, 1966.
10. Robert C. Cooke, *op cit.,* p. 70.
11. *Ibid.,* p. 71.
12. Otto Hué, *Krupp und die Arbeiterklasse,* Essen, 1912.
13. Josué de Castro, *The Black Book of Hunger, op. cit.* p. 6.
14. Josué de Castro, *The Geography of Hunger, op. cit.* p. 176.
15. G. B. Cressey, *Geographie Humaine,* Paris, 1939.
16. Josué de Castro, *The Geography of Hunger, op. cit.* p. 181.
17. Kenneth Boulding, *The Economics of Peace,* New York, 1945.
18. United Nations Food and Agriculture Organization, *The State of Food and Agriculture,* Rome, 1966.
19. Lord John Boyd-Orr, *Population, Food, Health, and Income,* London, 1936.
20. Josué de Castro, *The Geography of Hunger, op. cit.* pp. 160–164.
21. *Ibid.,* pp. 71—72.
22. J. R. Slonaker, *American Journal of Physiology,* Nos. 71, 96, 98, 132, 1925— 1928.
23. Romain Rolland, *L'Ame Enchantée (The Enchanted Soul, A World in Birth),* Paris, 1934.
24. Henry Reuss, Member, House of Representatives (D.-Wisconsin), Congress of the United States of America, 'Agricultural Development in Latin America,' a report prepared for the Committee on Banking and Currency, U.S. House of Representatives, Washington, D.C., February 1967.
25. Gunnar Myrdal, *An International Economy,* New York, Harper, 1956, pp. 230—231.
26. Stacy May, 'The Outlook for Industrial Raw Materials Demand in 1980 and Its Relation to Economic Development,' World Population Conference, Rome, 1954.
27. United Nations, Economic Affairs Department, *Relative Prices of Export and Imports in Underdeveloped Countries,* New York, 1949, p. 7.
28. *Ibid.,* p. 126.
29. UN General Assembly Resolution 623 (VII), adopted 21 Dec. 1952, *Official Records, Seventh Session, Supplement 20.*
30. United Nations, Economic Affairs Department, *Relative Prices of Exports and Imports in Under-Developed Countries,* p. 17.
31. Alice Mary Hilton, 'The Cybercultural Revolution,' *TMO,* October 1963.
32. *Idem.,* 'An Ethos for the Age of Cyberculture', *Proc. Jt. Computer Conf. AFIPS,* Spring 1964, London and New York, Macmillan, 1964.
33. Josué de Castro, *The Black Book of Hunger, op. cit.* pp. 7–8.

34. Harold Wilson, *The War on World Poverty.*
35. Josué de Castro, *The Black Book of Hunger, op. cit.* **p. 23.**
36. *Ibid.,* p. 16.
37. William Vogt, *Road to Survival,* New York, Wm. Sloan Associates, **1948.**
38. Gunnar Myrdal, *An International Economy, op. cit.,* pp. 299–300.
39. Lord John Boyd-Orr, *The White Man's Dilemma,* London, Allen & Unwin, 1953, pp. 99 ff.
40. Alice Mary Hilton, 'Cyberculture — Age of Abundance and Leisure,' *Michigan Review,* III, No. 4, pp. 217—229.
41. Lord John Boyd-Orr, *The White Man's Dilemma, op. cit.,* p. 102.
42. Josué de Castro, *The Black Book of Hunger,* p. 144.
43. Lord John Boyd-Orr, 'The Food Problem', *Scientific American* Vol. 183, No. 2, August 1950.
44. Josué de Castro, *The Geography of Hunger, op. cit.,* pp. 144–152.
45. Earl Parker Hanson, *New Worlds Emerging,* New York, 1949.
46. Barbara Ward, *The Rich Nations and the Poor Nations,* New York, W. W. Norton, 1952, p. 143.
47. Garnett de Bell (ed.), *The Environmental Handbook,* New York, Ballantine Books, 1970.
48. Alice Mary Hilton, *The Promises and Dangers of Technology,* New York, ICR Press, 1971, pp. 14—18.
49. Colin Hurry, personal communication.
50. Alice Mary Hilton, 'The Distribution of the Good Life,' *Proceedings,* Vth International Congress of AMIEVS, Havana, Cuba, April 1971.
51. Frederich Soddy, *Cartesian Economics,* London, Henderson's, 1922, pp. 18—19.

Against Pollution:
Chemicals

M. W. THRING

Physical and Chemical Aspects of Pollution

> ...when your fountain is choked up
> and polluted, the stream will not run
> long, or will not run clear with us,
> or perhaps with any nation.
> *(Edmund Burke 1729—1997)*

Statement of the problem: Can every human being have a life of quality in the twenty-first century?

In this paper I shall use the expression 'the quality of life' to express the way in which an individual human being feels that his or her life is worthwhile. This is, of course, entirely subjective and can never be measured on an agreed numerical scale. It is necessarily outside the scope of the physical sciences, but so are joy, happiness, love, creativity, and all the feelings that matter most directly to the individual. The physical scientist clearly cannot afford to make the mistake of saying that, because these things are outside the scope of his science, they do not exist or that they do not matter.

The quality of life requires freedom from negative things like hunger, cold, ill health, bad air and water, overcrowding, noise, stress, and fear, but it must also contain for every human being in the world the positive aspects of:

Frequent access to trees, birds, unspoilt countryside and seaside;
Beauty in all man-made surroundings and artistic creations;
The possibility of learning all one wants to learn, developing all one's motivations and talents to the full, and travelling to all parts of the world;
An opportunity to make a worthwhile contribution to the needs of other people both in one's paid job and as a hobby.

Obviously, different people would define the aspects of a life of quality differently, but everyone can agree that the expression 'the quality of life' means something approaching those things most important to the individual. While no two people can agree on exactly what they mean, there are many statistical measures of the deterioration of the quality of life in large groups of people. An increase in suicides, alcoholism, drug use, and such stress-linked ailments as heart, digestive, and skin ailments, asthma, migraine, arthritis, and a wide range of psychological disturbances, provides statistical evidence that the quality of life in a given city is falling. These, then, must be direct consequences of the side-effects or by-products of the use of machines to give us an affluent society.

The problem we have to investigate is, therefore: How can we give every human being in a world, which will probably contain 7×10^9 people by the year 2000, the things essential to a life of quality without subjecting them to the poisons which destroy it? I shall use the word 'poison' to describe any pollutant which, in sufficient concentration, can cause severely harmful physiological or psychological effects, directly or indirectly, on humans. Such effects may be killing them or harming their descendants, destroying their enjoyment of life, or driving them mad. This includes the indirect effects of such things as drabness and ugliness, excessive illumination and garish colours, the destruction of vegetation and birds, as well as the more direct effects of noise, stress, and food, water, and air poisoning.

There is a large margin of scientific uncertainty about the physiological effects of small doses of poisons in the air, water, or food of man. This margin of uncertainty arises from the fact that the legal threshold of safety must be set at points where scientifically measurable damage to human health can be shown to be at a high level of statistical significance. These legal thresholds are primarily established for medically checked and healthy adult workers, exposed for a fraction of their lives – at most 80,000 hours, if they are exposed during their entire working lives of forty hours per week for forty years, which, out of a lifespan of more than 640,000 hours, is one-eighth. The thresholds, however, are clearly inappropriate when applied to people exposed to these poisons from birth to death and born of parents who have also had a lifetime of exposure.

In almost every case in which the general public is subjected to poisons, without any possible choice of escape, the level of amplitude of poison concentration is an order of magnitude below the legal threshold. It is, then, extremely difficult, if not impossible, to produce scientific evidence of medical or psychological damage, especially of long-term effects. The noise from traffic or aeroplanes may be two orders of magnitude (20 decibels) below the level known to cause deafness, and yet it certainly causes disturbance, irritation, and loss of sleep. The maximum concentration of carbon monoxide in city streets rarely exceeds one-tenth of that proved to be harmful to industrial workers, and yet everyone knows how one longs to get away from the stale air of these streets.

It is this wide gap — between the level of poison which can be clearly proved to be poisonous and the level which has an insidious, long-term, and yet definitely harmful effect on the quality of life — which allows those whose economic interest it is to emit the poisons to reject as unscientific the arguments of those who are upset by their observations of these poisons. It is difficult, if not impossible, for laws to be passed prohibiting poisons at levels too low to be accurately measurable. Yet anyone can observe for himself that these poisons have an immediate effect on the quality of his life. The effects, furthermore, can be shown to build up steadily, in many cases decade by decade.

There is another major problem in this field which is summed up in the old expression 'one man's meat is another man's poison,' and that applies to all low intensities or concentrations of poison — the threshold of unpleasantness varies enormously with the individual: the more sensitive can be prevented from intellectual concentration and even driven to disraction by a noise or upset by a smell or taste that another person may barely notice. Headaches, digestive upsets, and nervous symptoms can be built up in a few hours' exposure in some people while less sensitive people can't see what the fuss is about.

These two effects — the gap between the sensitivity threshold and the statistically harmful threshold and the variability of the sensitivity threshold — are the reasons why the method of medical statistics is far too coarse a net to catch the finely dispersed poisons steadily eroding human enjoyment of life, especially in the big cities of a highly developed, affluent society. It is through this gap that the emitters of poisons can escape the responsibility for their actions. Moreover, in making a law it is exceedingly difficult to prove a poison's existence in terms of its concentration at every point in the environment and of the long-term buildup. Laws are usually formulated in terms of the concentration at or close to the emitter. Thus the answer to the question posed in the heading of this section: 'Can every human being have a life of quality in the twenty-first century?' is 'No, not unless we can succeed in producing a complete change in the objectives and direction of progress of civilization.'

In order to separate the basic factors in pollution, we can write an equation for a region of the world which can be regarded as practically closed, i.e., as retaining the bulk of a given pollutant emitted within the region itself. We shall see later that a Megalopolis can be regarded as such a region for some pollutants, but that in other cases it is the whole northern hemisphere that is a region.

Concentration of a pollutant in the region

$$\infty \frac{n}{A} \cdot G_{c+w} \cdot t \cdot C_f$$

where n = the number of people in the region, A = the area of the region, t = the halflife of pollutants in the region and in people in years, G_{c+w} = the goods consumed and wasted per head per annum, and C = the carelessness factor which has a maximum value of 1 if no trouble is taken to reduce the production or emission of the pollutant into the enviromment.

In the concluding section of this paper, I shall attempt to put forward some positive, concrete proposals as to how mankind, in its attempt to acquire a high standard of living for everyone alive on a crowded planet, can avoid poisoning itself to the point where the quality of life becomes unbearably low. First, however, I will try to summarize our knowledge of the quantity of poisons emitted, what happens to them after emission, how harmful they are in various concentrations, and which are the best ways of reducing pollutants without reducing too drastically the useful results of which pollutants are unwanted by-products.

Pollution and pollutants

I have, for convenience, divided pollution into the three headings air, water, and land, even though many problems such as radioactive atoms or organic pesticides are common to all three.

Air pollution

Air pollution, noise, ugliness, and excessive light intensity are all poisons in the sense defined above, because they can have severely harmful physiological and psychological effects on humans. I propose, however, to confine myself to a consideration of gases, dusts, and thermal effects on the atmosphere.

The following is a list of most of the pollutants known to be emitted into the atmosphere in substantial quantities as an unintentional result of technological activities.

Gases. Products of complete combustion of hydrocarbons, CO_2 and H_2O constitute 90 per cent of the total mass emitted into the atmosphere. These are not normally described as pollutants since they also play a large role in natural processes and are not in themselves poisons. (Several of the remaining gases also play a role in natural cycles, but in each case man has on occasion misused their local concentration to dangerous levels.)

Products of incomplete combustion or thermal decomposition of hydrocarbons, fats, and glycerol, CO, $C_n H_m$, benzpyrene, aldehydes, partially oxidized hydrocarbons.

Sulphur compounds, SO_2, SO_3, H_2SO_4, and H_2S (mainly from stationary combustion of high sulphur fuels).

Oxides of nitrogen NO_x, especially NO and NO_2.

Chlorine and fluorine compounds, especially HCl and HF.

Arsines.

Hydrogen cyanide (fumigation, blast furnaces, metal plating).

Phosgenes (chemicals and dyes).

Radioactive gases.

Ammonia.

Particulates and aerosols. These constitute about 10 per cent of emissions by weight.

Iron oxide.

Cement and lime dust.

Pesticides.

Pb compounds. (It is estimated that more than 10 mg/m^2 average has been deposited over the land area of the USA from car exhausts since 1923.)

Other metallic compounds, such as Hg, Ba, Sb, Bi, B, Cd, Cr, Co, Cu, Mn, Mo, Ni, Se, Sn, Ti, V, Zn.

Asbestos.

Soot.

Fly ash.

Radioactive dusts.

There is no danger of man's seriously depleting the oxygen content of the atmosphere — if we were to burn all the known fossil fuel reserves we would use less than 3 per cent of the available oxygen.[1] The fear has been expressed that the herbicides and pesticides which find their way into the sea might upset the basic photosynthetic organisms, while the replacement of green vegetation on the land by concrete will reduce the land photosynthesis. But 0.05 per cent of the O_2 in the atmosphere is renewed each year by the photosynthesis, 60 per cent of this being in the oceans.[2] There is no evidence, however, of any lowering of the O_2 of the atmosphere or in the sea as a whole, and the real problems relate to the food supply, the lowering of O_2 in lakes and rivers, the presence of CO in local regions, and perhaps eventually to the CO_2 content of the atmosphere. It has been estimated that 5×10^9 tons of CO_2 are added annually to the atmosphere at the present rate of world fuel consumption.[3] This would be enough to increase the CO_2 in the air by 2-3 ppm if the CO_2 were uniformly distributed and not removed.[4] In the last hundred years the CO_2 content of the atmosphere has risen from 290 ppm to 322 ppm, the rise from 1959 to 1969 being from 313 to 320 ppm. This increase accounts for nearly half of the CO_2 released by combustion of fossil fuels.[5] The remainder must have been reabsorbed mainly by the oceans, but land vegetation has also probably increased to some extent. Between 1900 and 1945 global temperatures rose at a rate of 0.008°C/annum and since then there has been a cooling effect at a lower rate. The rise may have been due to the rise of CO_2, reducing the escape of the infrared reradiation from the earth's surface. It has been calculated that if the world's thermal power output in the year 2000 is six times that of the present this waste heat will not affect the global atmospheric temperature directly except near cities ('heat islands'), where it will increase the heat reaching the ground by 10 per cent. On the other hand, if this waste heat were produced by fossil fuels there would be an 18 per-cent increase in the CO_2 content of the atmosphere, and this could possibly raise the earth's temperature by 0.5°C.

Particles in the troposphere will absorb and scatter the sun's radiation and the infrared reradiation from the earth and also affect the properties and amount of cloud. They could probably produce a cooling effect in proportion to their concentration, but it is unlikely that the global effect of man-produced particles is significant. On the other hand, the natural Agung eruption produced a thirty-fold increase in stratosphere particles and an increase of 6–7°C in the stratosphere temperature and a small decrease of global surface temperature.[6]

Jet aircraft in the upper troposphere produce condensation trails which may upset the radiation balance slightly, and might initiate precipitation and cloud formation by falling ice crystals. (Deliberate cloud-seeding experiments with massive doses of AgI nuclei have caused seeded clouds to rain three times as much as (control) unseeded clouds.[7])

It is almost impossible to pinpoint any particular air pollution as the main factor in causing human deaths and ill-health although arsenic asbestos and coal tar have been proved to be carcinogens while SO_2 and CO are known health hazards. However, most of the medical statistics can only compare generally polluted with unpolluted air. When Lave and Seskin studied the long-term effects of living in a polluted atmosphere, they obtained statistics from 53 boroughs and related mortality rates to a solids deposition index.[8] Thirty-nine per cent of the bronchitis mortality rate was explained by the regression – a ten per-cent decrease in the deposit rate would lead to a 7 per-cent decrease in the

68

bronchitis death rate: a smoke index gives an even better correlation — cleaning the air to the best value would reduce the bronchitis mortality by 70 per cent. Similarly, a regression on lung cancer in England and Wales showed that if the quality of air in all boroughs were improved to that of the best, death rates by lung cancer would fall by between 11 and 44 per cent. Rates of mortality in part of New York State due to stomach cancer are more than twice as great in areas of high pollution as in areas of low pollution. London postmen have 25–50 per cent higher rates of occurrence of severe respiratory disease than small town postmen, while 20–35 per cent of absence due to sickness in London bus drivers could be ascribed to air pollution measured by a fog index. Infant death rates in Nashville showed the highest correlation with atmospheric concentrations of SO_3. Lave and Suskin conclude that 4.5 per cent of all economic costs associated with morbidity and mortality vould be saved by a 50 per-cent reduction in air pollution in major urban areas.

What happens to air-borne emissions? When a car emits exhaust a few centimetres above the ground, the most important pollution problem is that in its immediate neighbourhood, viz. 10–250 m away and 0–2 m in height. This is especially true in traffic jams, tunnels, and streets hemmed in by high buildings. It is, therefore, convenient from the point of view of dilution, dispersion, and disposal of poisons to divide the atmosphere into three zones as follows.

Zone I comprises the immediate neighbourhood of the source: up to 250 m radius and 10 m high, the height of a 2–3 storey building. In this region occur concentrations of poisons up to 100 ppm (0.01 per cent) because the concentration in the exhaust gases and chimney gases may be of the order of 1 per cent and even 10 per cent in the case of idling car exhausts, and the exterior mixing can be as little as 1000-fold. The time scale for dispersion in this region is of the order of one hour. The height of the source above ground level is extremely important in this region because of the gradient of wind velocity with height and because the distance to the worst concentration at ground level is roughly proportional to the height. Bosanquet gives distance to maximum concentration 10 (H + h),[9] and the peak concentration at ground level is inversely proportional to the square of the effective height. The effective height is the actual height (H) plus extra height (h) due to the buoyancy of the gases if they are lighter than the surrounding air. Both these effects are most important in very light winds. The buoyancy effect can be very valuable if the gases are much hotter than the air, so wet washing may be undesirable from the point of view of local effects; combining several gas flows into one chimney is very valuable from this point of view. In an inversion, the temperature of a grey body such as a particle-laden chimney gas can fall below that of the surrounding, more transparent air giving a fall to the ground without much dispersion. It follows that if vehicles had exhausts emitting the gases vertically from a height of 2 m above the ground, the concentration of CO and other poisons would be greatly reduced especially at heights below 2 m. This would be of particular value in a traffic jam. It would also make the emission of visible smoke much more conspicuous.

Zone II corresponds to a whole valley or town — a region of 1–100 km² with a time scale of the order of one week so that poisons accumulate seriously if there is inversion and lack of wind lasting some days. The Californian inversions are the classic examples of a region which accumulates its own poisons. The weekly average level of CO in the Los Angeles basin can be as high as 20 ppm.[10] Here the dilutions are of the order of 1 in 10^7 compared with the concentrations emitted from chimneys and car exhausts, but these have been shown to cause certain harmful reactions, e.g. SO_2 is oxidized to SO_3, NO_2 produces O_3 far below the normal zone layers, and trace hydrocarbons contribute to a lachrymatory effect while CO can assist the oxidation of NO to NO_2.[11]

Well-known examples of serious buildup of poisons in Zone II in Europe have been the Meuse Valley sulphur damage in 1930[12] and the London smog of December 1952, which has been indicated to have caused 4000 premature deaths by bronchitis. However, Garnett indicated that even during the inversion period of the London smog, 96 percent of the SO_2 was removed by meteorological mixing from the atmosphere mainly at night by droplets falling to the ground.[13]

It is significant that in the book *Air Pollution,* based on a conference held in Sheffield in 1956, only Zones I and II were considered, whereas in the fifteen years since then the increase in industrialization and especially fuel consumption in factories and vehicles has made it necessary to consider also Zone III, so that high chimneys are no longer a satisfactory way of dumping poisons such as SO_2.

Zone III covers areas ranging in size from 100 to 10,000 km², and thus, ultimately, comprises, for example, the whole northern hemisphere right up to the stratosphere. Here the time scale for dispersion is of the order of one year. One can regard the northern and southern hemispheres as separate regions to a first approximation. Because the main bulk of the fuel consumption is in the northern hemisphere, it has been found that near sea level there is a boundary close to the equator with higher CO levels (0.20 ppm) on the north side: On the south it is about half at 0.10 ppm.[14] However, the world total of 200 million tons of CO per year would correspond to an increase of 0.03 ppm per

70

year if mixed through the whole atmosphere: such an increase has not been found, and the concentration decreases in the stratosphere so that it may be oxidized to CO_2 ($CO + [OH] \rightarrow CO_2 + [H]$) above the tropopause. Inman, Ingersoll, and Levy conclude that the CO is absorbed in the soil by micro-organism metabolism at 8.44 mg/hr–m^2.[15]

A recent calculation indicates that the sulphur emission from chimneys in Britain and Central Europe is likely to be transported by air movement to Scandinavia during periods of dry weather so as to contribute a major part of the observed concentrations of sulphur.[16] These calculations are based on the assumption that fuel containing 2.5 per cent sulphur is burnt to the known amount in fully mixed boxes 2° in latitude and 2° in longitude and carried by the known winds at 1.5 km height.

Vehicle emissions. In the Table below our present knowledge of the poisons emitted by the internal combustion engine is summarized. Both the second and third columns must be multiplied by the fuel consumption rate per hour to obtain a true comparison between two vehicles. The first and most obvious way to reduce emission is either to use less horsepower-hours to obtain the required transport by smaller engines or to have more people in one vehicle.

Until recently most attention has been paid to this subject in the USA, where the large private automobile is most widely used. Due to the special climatic conditions in Los Angeles, the problem is particulary severe there. Senator Muskie's amendment to the Clean Air Act in December 1970 requires a 90 per-cent reduction in CO and $C_n H_m$ from 1970 to 1975 and in NO_x from 1971 to 1976.[17] Attention is now being directed to the same problems in Japan and Europe.[18][19]

There is absolutely no conclusive evidence of the danger to health from CO, hydrocarbons, soot, lead, or oxides of nitrogen emitted by cars, but there is little doubt that in both Zones I and II several of these are real poisons in the sense defined earlier in this paper. The air in city streets causes depression and headaches wherever traffic jams occur, while at least half of the well-documented poisonous results in the Los Angeles area are attributed to vehicles. Lead alkyls have been put into petrol as a knock inhibitor for fifty years, at the rate of 2g/gal in regular and 4g/gal in premium fuel (1 gal = 3.5 kg). As a result, lead is undoubtedly accumulating to a dangerous degree in the environment. Murozimi et al. showed that from 1750 to 1940 the Pb content of the snow layers in northern Greenland increased from 0.01 to 0.07μg/kg, but that since 1940 it has been rising at the rate of

Motor vehicle exhaust emissions[56,59]

Emission	Maximum per cent of exhaust	Weight per cent of fuel	Total weight per year (in 10^6 tons)		Maximum ppm in atmosphere	Notes
			GB	USA		
CO	10	25 (15 % fuel c.v.)	5	58	360	Highly toxic, 50 ppm TLV — 2 hrs
C_nH_m	900 ppm	6 (+ 2 % crankcase; 1.5 % evaporation; 15 % fuel c.v. wasted)	0.25	15.1	15 (5)	500 ppm safe limit
Odour	—	—	—	—	—	Smell unpleasant, carcinogens?
Soot	—	?	—	1.1	—	Dirt and fog, vehicles 30% of the total in USA; diesels
SO_2	—	—	—	0.7 (9)	—	Vehicles 1% of total
Pb	—	0.1 (7)	—	0.18 (7)	44 µg/m^3	200 µ/gm/m^3 TLV
NO_x	1600 ppm	2	—	0.17	1.5 ppm (5)	Power Plants = ½ cars, 3 ppm for 1 hr is serious

0.12 μg/kg every year. (Pb has been found at the rate of twice the internationally recommended safe limit in 3 out of 47 samples of public water supplies.[20]) The average blood level of Pb in a group of children in Manchester is 0.30 ppm, the average for US adults was measured at 0.26 ppm; the known level at which enzyme inhibition by inorganic lead occurs is 0.2–0.4 ppm.[21] Pb ingestion is known to cause brain damage, ranging from irritability to mental retardation in children. The possible cures can be listed as follows (excluding raising the exhaust pipes of automobiles to 2 m above ground level to accelerate short-term Zone I dispersal and using fewer horsepower-hours):

Treatment of exhaust gases. High- and low-temperature catalytic afterburners can deal with CO and other products of incomplete combustion, but their use tends to lower overall efficiency and to increase weight, thus increasing fuel consumption. Catalysts are poisoned by Pb so that catalytic afterburners require a non-leaded fuel. Exhaust gas recirculation can reduce NO_x formation by lowering peak combustion temperatures. A 75 per-cent reduction to 400 ppm is produced at part load operation by 15 per-cent recirculation, but the peak HP of the engine is correspondingly reduced.

Use of different fuels in present type of engines. High-octane fuel without Pb can be made from conventional petroleum by considerably more expensive refining and reforming with a lower yield. LPG (mainly propane) must be stored under slight pressure, but can be used in gaseous form so that the problems of accurate fuel/air proportioning to all cylinders at all times are greatly simplified. However, a very small fraction of the total petroleum can be converted to LPG so that world petroleum resources would be rapidly exhausted if this became the main vehicle fuel. CH_4 (natural gas) has been used in the south of France for many years and recently in some experimental vehicles used in the USA. This fuel is readily available and excellent, but being a gaseous fuel it must be carried at very high pressures in heavy steel bottles. If balloons are used, fuel can be carried for only a short distance.

Modifications to the conventional piston engine. Optimum proportioning of fuel and air at all times and to all cylinders and optimum ignition timing, such as vaporizing the petrol before mixing, low pressure injection of exactly metered quantities of liquid fuel into the manifold, the use of a separate carburettor for every cylinder have all been proposed and tried. This has the clear advantage of reducing the total fuel consumption simultaneously with the unburnt material. It is also possible to reduce the NO_x formation by going well to the lean side of stoichiometric so that the reduction of temperature more than offsets the extra O_2 available, but this means a reduction in maximum power because less fuel can be burnt.

73

High-pressure fuel injection into the cylinder as in a diesel engine does not require Pb or high-octane fuel, but requires a troublesome and expensive pump, is more rough running and liable to emit soot and small percentages of unburnt and smelly hydrocarbons, although these can be avoided by good design and maintenance. The stratified charge engine is being developed to give very good combustion with excess air of any fuel (i.e., no Pb) by controlled mixing of injected fuel with the air in the cylinder.[22]

Hot-wall engine. Since much of the problem of unburnt material in exhaust hydrocarbons is due to quenching of combustion of fuel by the cooled walls of cylinder and piston, some experiments have been made with a cylinder and piston of SiN, which has a lower thermal conductivity than cast iron and can be run at a much higher temperature. This again can reduce fuel consumption.

Radically different power systems. Of the rotary intermittent combustion engines, the most highly developed is the Wankel engine, which does not emit any less unburnt material than the conventional engine because of cooling effects. But its smaller size leaves more room for afterburners.[23] If a rotary engine can be developed with hot walls and good combustion chamber shape, this could give a great improvement in the fuel consumption as well as eliminating unburnt material.

Hybrid systems may be developed. One of the main reasons why the internal combustion engine causes pollution by incomplete combustion is the sudden movement of the accelerator required in traffic. A constant-speed engine which charges a large battery or accelerates a very-high-speed flywheel (carbon fibres have been suggested) or compresses air in a very-high-pressure bottle as a power storage for a hydraulic system could all be used. In all these cases the maximum power of the engine need be only the time-*mean* power requirement, which is of the order of a fourth of the maximum so that the engine can be much smaller as well as running always at the optimum speed. In Germany a hybrid diesel/electric bus has been operated which can run for three miles on the battery alone in the most heavily polluted part of the town. Even better results could be obtained by combining a continuous-combustion engine (see below) with an electric battery. This is, indeed, the only system known at present which is likely to meet the proposed 1980 US regulations regarding completeness of combustion and oxides of nitrogen.

The all-electric car in which the lightest type of Pb-acid battery could be used would serve for a town car with a range of 30 miles and maximum speed of 30 mph, but it would still be very heavy. So far, although much work is being done, no lighter battery has been successfully developed which is reasonably cheap and will stand hundreds

74

of recharges without deterioration. A fuel cell which consumes a readily available fuel with air is feasible, but most present types must operate at medium (200°C) or high temperatures (700°C). There is no reason why such a fuel cell should not provide an adequate non-polluting power source for a town car in a few years, and the heating could be done by a small, continuous combustion system.

It is very much easier to obtain complete combustion and avoid oxides of nitrogen in a continuous combustion system than in the difficult conditions of millisecond combustion time, rapidly changing temperature and pressure, and cold walls of the piston internal combustion engine. Moreover, it is quite easy to make a continuous combustion system burn a wide range of liquid gaseous, or even solid fuels. Three systems are already available: the steam engine, the gas turbine, and the Stirling engine. The steam engine requires a very large condenser and has frost problems, but is particularly interesting in conjunction with the hybrid electric system (see above) and could undoubtedly make great progress if a sum of money comparable with that spent on the internal combustion engine were spent on research. New working fluids with better properties than water or even binary systems (water + low-temperature fluid to reduce condenser size) new ideas for once-through miniature heat exchangers and a very-low-fuel consumption all-night burner to prevent freeze-ups could all contribute greatly, especially if we change our attitude to the car from a highpower status symbol to a transport device.

The gas turbine with a heat exchanger can give high efficiencies and low pollution, but is unlikely to be practical below several hundred horsepower of a large lorry because of the turbine blade manufacturing costs and the scale effect on efficiency. The Stirling engine has already been developed to a point where it could be used on cars, and further work will probably reduce the production cost sufficiently to make it a fully feasible proposition.

Pollution from chimneys and factories

Until about 1950, smoke from incomplete combustion of coal was the most noticeable poison emitted from factory and domestic chimneys and even power stations on occasion. Grit and fly ash from large boilers also gave serious ill effects. In Britain, the Clean Air Act of 1956 has reduced these problems to secondary importance by the creation of smokeless zones in which tarry fuels may not be used and of enforceable restrictions on visible emission everywhere else. Partly because solid fuel is not significantly cheaper than oil or gas the proportion of domestic heating done by solid fuel has dropped greatly.

In the cases of both smoke (soot) and CO, the elimination of pollution is purely a matter of good housekeeping, i.e. the proper running and maintenance of proper equipment, because the fuel costs are reduced by the same improvement of combustion. It is moreover fairly easy to remove coarse grit ($> 200\mu$) and dust ($200-30\mu$) so that the worst poisons now are oxides of sulphur and fine dusts and fumes (iron oxides, cement, other metal oxides). Oxides of nitrogen can be increased 500 times compared to the natural levels, but there is as yet no evidence that these matter except in the case of an isolated Zone II caused by persistent inversion. HCl can be emitted in significant quantities when a chlorine-containing fuel or plastic is burnt or when NaCl enters the flame, while HF has given considerable trouble in the neighbourhood of aluminium smelters but can be washed out without excessive cost burden. SO_2 has been found in concentrations as high as 5 ppm around a sulphuric acid plant[24] but only up to 0.3 ppm around power stations, while the British export to Norway, according to Reiquaim's calculation, could reach 12 μg/m^3 (0.004 ppm).[25] In London, during the winter inversion of 1962, the concentration reached 1.4 ppm, while the average concentration throughout the UK is 0.04 ppm.[26] In the atmosphere, SO_2 is slowly oxidized to SO_3, and so forms sulphuric acid in condensate drops. Sulphuric acid mist can cause irritation to sensitive humans at 0.1 ppm.[27] There is a clear correlation between SO_2 pollution and the bronchitis death rate.[28] Thus there is little doubt that chimney emissions of SO_2 caused by burning oil and coal with more than 1 per cent S by weight are major poisons.

This SO_2 can be reduced by 90 per cent in continuous operation either by desulphurizing the fuel or by wet or dry cleaning the gases after combustion. A short-term temporary measure which has been used is to burn low sulphur fuel only at times of inversion. The long-term solution for domestic and small-scale industrial heating is to use low sulphur fuels. Natural gas and town gas are either sulphur-free or readily desulphurized, and it is not too expensive to reduce the content of light distillate oils to 0.3 per cent. The alternative is to go in for a combined power station and pass-out steam district-heating scheme, as has been done in Moscow. This enables one to apply efficient cleaning methods to the large power station.

A strong plea has been made by Squires that coal engineers should have money, as nuclear engineers have had, to pursue long shots and particularly to study means of pretreating coal so that its sulphur is extracted.[29] This is an important line of research from the point of view of not putting all our power-generation eggs in the nuclear basket and the very limited world supplies of petroleum.[30] Fifteen years ago, my department at Sheffield did some work on the use of dolomite to remove SO_2 from combustion gases. NAPCA is now supporting further work on this subject, using a fluidized bed. Whether it is the wet methods (with their disadvantage of reducing the buoyancy of the gases but with a potentially higher degree of SO_2 removal) or the dry methods that are eventually used, depends on the degree that more intensive development can improve them. Certainly, we can no longer rely on the use of high chimneys (already up to 300 m, and one of nearly 400 m is being built in Ontario) to throw the SO_2 farther away. New Jersey has already issued rules calling for SO_2 emissions equivalent to no more than 0.2 per cent in coal or 0.3 per cent in oil.

In refineries there are problems, with smells resulting from very low concentrations of certain vapours, which can be largely dealt with by using floating tank roofs (to avoid vapour breathing to the atmosphere), deodorizing effluent water, and burning any vapours. Much work has been done on complete incineration of solid wastes from towns, but here again there is a strong case for large central systems doing magnetic separation and air elutriation and then burning very efficiently to make steam and cleaning the effluent of HCl, SO_2, and all other poisonous constituents.

The removal of fine dusts and fumes with efficiencies up to 99 per cent is quite feasible for nearly cold gases using well-known equipment such as wet electrostatic precipitators, fabric filters, and venturi scrubbers, and such equipment is already in use when the clean gas or the particles removed have sufficient economic value. I have proposed a combined regenerative heat exchanger and thermal precipitator for glass tanks. In general, there is no doubt that if really clean stack gases were required (for the local amenity or for possible climatic effects) they could be produced at a cost increase of the product of the order of 10 per cent. This would be done by installing available equipment and by spending more on trying out a range of new possibilities.

Water pollution

The per capita use of water in the USA is doubling every 40 years.[31] Sewage alone uses about 150 gallons per capita per day. We know how to treat sewage with activated sludge to remove suspended solids and biological oxygen demand (BOD), but at the moment only just over half the US sewage is treated by the best available methods. In England, seaside towns discharge directly into the sea. Experiments are under way for using O_2 instead of air in the activated sludge process and for physical-chemical methods, e.g. lime precipitation and carbon adsorption.[32] Secondary biological treatment such as activated sludge normally removes only about 30 per cent of the phosphates in sewage, which come in large quantities from detergents, but lime precipitation can be more effective. Much research is going on to find efficient methods of nitrogen removal, but a completely satisfactory system is not yet available.

The sludge produced is expensive to handle as it must be dewatered — by cooking or centrifuging, and the solids disposed of as land fill or sold as fertilizer. It is certain that in the future the recycling of the solids for agricultural use to replace humus in the soil must become the conservation system. Thus we have a long way to go before we can make the fullest use of our limited fresh water reserves and of our organic refuse.

The process of eutrophication or overfertilization of natural bodies of water, such as ponds, lakes, and partially enclosed coastal marine water, is a major problem, for example in Lake Erie and off Long Island. Algae and other plants grow so profusely that the BOD of the water rises sufficiently to kill all fishes and other aquatic animals. It may take a century for the process to be reversed. Nitrogen and phosphorus are both causes of this fertilization of the algae and come from detergents and the runoff from heavily fertilized agricultural land. Ryther and Dunstan have shown that the coastal waters receiving the pollution from New York harbour have more than sufficient phosphate but are short of available nitrogen so that the moves to substitute nitrilotriacetic acid for phosphates in detergents may, in fact, increase eutrophication in these cases.[33]

The output of industrial waste water is some 2.6 times as much as that of domestic waste in the USA. In Britain we have been throwing industrial effluent into the river Mersey for a hundred years and neither fish nor algae can live in it. Phenols, cyanides, sulphate, manganese, nitrate and nitrite, choride, fluoride, copper, lead, zinc, chromium, sulphide, mercury, complex organic chemicals, and arsenic are among the pollutants for which a check has to be kept in river or sea water into which factories are discharging their effluents.

Arsenic is a cumulative poison which builds up slowly in the body. The maximum allowable in drinking water, as recommended by the US Public Health Service, is 10 ppb. Angino et al. have found up to 8 ppm in the Kansas River and up to 41 ppm in certain detergents and pre-soaks.[34] Arsenicals are also used as farm insecticides.

Tolley and Reed of the University of Liverpool have sampled the tap water of 43 English boroughs and found that 13 per cent were contaminated with lead to more than 0.10 mg/1, at which figure the World Health Organization recommends that water should be rejected for public supply.[35]

Twenty years ago, fishermen, their families, cats, and the local sea birds at Minamata Bay in Japan suffered severe structural damage to the brain which was traced to alkyl mercury compounds discharged by a factory into the sea concentrated by a factor of more than 50,000 in the fish. Farmers who have themselves eaten grain seed treated with organic mercurial fungicides or have fed it to their pigs have had severe poisoning.[36] Alkyl mercury can cause congenital mental retardation. It is quite certain that great care must be taken whenever there is any possibily of mercury compounds reaching fishes or animals which will eventually be eaten by man. Goldwater recommends that toxic mercurials in industry and agriculture should be replaced by less toxic substitutes.[37] To this must surely be added that the whole conception of towns and industries using lakes and rivers as dumps to take waste material away is completely incompatible with man's ever reaching equilibrium with his environment. If biological processes can concentrate certain inorganic or organic chemicals by huge factors, anyone wishing to dispose of any chemical whatever must either refrain from diluting it or find a way of concentrating it so that it is available as a useful raw material and not allow it to enter the exterior enviromment in any measurable quantity.

Fluoride is put into drinking water in many places at a level of 1 ppm to reduce dental caries.[38] It is suggested by Cook that there may be some evidence that even at this level enamel hypoplasia occurs in 10—15 per cent of children.[39] In high-fluoride areas (up to 4 ppm) for example in North Africa, it has been reported that 'darmous' occurs and the teeth fall out or wear down to the gums.[40] The same apparently happens to cattle exposed to heavy quantities of fluorides. Resorption and exchange of calcium in bone are inhibited by fluoride and, if there is a deficiency of calcium or vitamins C or D, the body is no longer able to deal with the fluoride.[41] One must, therefore, consider that the deliberate fluoridation of water may not be a safe long-term practice.

Man has always treated the seas and oceans as a self-cleaning sewer for all kinds of ships' refuse, liquid factory effluent, and seaboard towns' sewage, but there are now many warning signs that this careless behaviour is also damaging the natural life of the salt water. Thor Heyerdahl has reported on the accumulation of flotsam from human waste covering vast areas of the ocean; Cousteau has stated that the percentage of lead in the top 100 m of the sea is five times what it was fifty years ago and that even the coral reefs are dying as a result of human refuse. Untreated sewage has caused shellfish to be found containing hepatitis, polio virus, and other pathogens. Abelson recommends that we need an ocean early-warning system for heavy metals, chlorinated organic compounds, and possible carcinogenic com-

79

ponents of petroleum by more intensive monitoring of algae and fish.[42]

By far the most spectacular poison put into the oceans is oil from damaged tankers, and ships cleaning their tanks while at sea, and even from coastal spillage from land tanks. Many of the best bathing beaches from Britain to Bondi Beach in Australia are frequently rendered completely useless by the tarry oil/water emulsions that float ashore. Birds are killed especially by the volatile aromatic constituents. The detergent emulsifiers used for beach cleaning in Britain after the Torrey Canyon spillage were toxic to marine life at 5 ppm. An oil pollution research unit has been set up in Britain to study methods of dealing with spills without so much damage to ocean life, but clearly the solution is to reduce the amount of oil spilled. The design of shore settling tanks to deal with the material cleaned from the ships tanks has been available for thirty years, but they are rarely used because shipping companies find it cheaper to pump the cleanings into the sea during the return journey.

'Thermal pollution', the heating of rivers and lakes and even local regions of the sea by the waste heat from electricity generation, is a major problem in many places and will become worse if we do not reach a steady-state level of electricity consumption per capita in the developed countries. The cooling tower is an alternative, but a water cooling system is cheaper. In the equilibrium system, however, all the waste heat will be put to a useful purpose.

Land and food

The gradual increase in the proportion of all kinds of poisons in human food and drink is detectable by modern methods of analysis. If we continue to allow organic compounds of heavy metals, pesticides, antibiotics, chemical fertilisers, and even disease-causing bacteria to enter the environment in ever-increasing quantities, there must inevitably be a steadily increasing deterioration in the quality of human life. It is difficult, if not impossible, to prove any deterioration in the taste of food, but there can be no doubt that these poisons are one cause of the present unrest in the developed countries as exemplified by drug-taking and stress illnesses.

Beaconsfield describes the three kinds of internal pollution: 1) Ingestion and inhalation of the products of our already polluted external environment; 2) Our daily intake of chemical additives and impurities pre-packed into our food and drink; and 3) The vast number of medicaments with which we regularly dose ourselves. He quotes an estimate that half a million Englishmen go to bed with sleeping pills, three times as many of their wives take slimming pills, and many of their children are 'purged' regularly.[43]

Fifty years ago public health authorities in developed countries were concerned about deficiencies of minerals and vitamins; now they are worried about the presence of trace amounts of pesticides, fertilizers, preservatives, radioactive fission products of fallout, and carcinogens, e.g., polynuclear hydrocarbons.[44] DDT has a halflife of half a year in the human body and about seven years in the ecosystem.[45] According to the amount put into the environment from which our food comes, therefore, we build up a corresponding equilibrium level. The heavy metals such as lead stay in our bodies so much longer that they must be regarded as cumulative poisons.

It may be a controversial matter whether factory farming in which hens, pigs, and cattle are permanently cooped up in crowded buildings is an ethical way to treat them, but there is some evidence that the water content of the meat is rising as a result, and many people feel that such complete degradation of the quality of life of the animals man feeds on must contribute to the degradation of quality of life of the feeder. Certainly this crowding of animals necessitates the routine use of antibiotics in the feed and these act partly as growth promoters. This continuous dosage causes the organisms in the gut to build up resistance to the antibiotic, and this causes a danger that resistant bacteria can attack both the animal and man and be untreatable by all known antibiotics. Moreover, minute traces of the antibiotics remain in the carcases and could have serious effects on a person who is allergic to the antibiotics used. The British Swann Report has recommended that no antibiotic used in animal or human therapy should be allowed in animal fodder.[46]

Darling refers to the danger of single crops. The English Downs, for example, have lost their ten-inch layer of humus. He also points out the effect of barley cropping until the organic content of the soil falls to 1.5 percent. The converse situation in the west of England, where cows are kept, is that there is so much animal manure that a sewage problem exists. Darling also describes intentional and careless dumping of unwanted hardware and litter as pollution of the land.[47] We suffer very badly, in England, from the mess made by nineteenth-century industry, which nobody considers his business to clear up.

The cutting down of trees and of the English hedges to increase the useful output of land is also a result of the economic pressure on the farmer and the higher wages in the factories which contribute significantly to the loss of the therapeutic effect of countryside on the human, if he finds time to walk.

DDT and other persistent pesticides have been an essential element in the production of cheap, high-quality food in the USA, with only 5 per cent of the labour force in the last twenty years. Dickinson comments: 'But the burgeoning degradation of our national and social environment indicates that perhaps we are paying too little for food that is unnecessarily free of blemishes, that we have driven more people off the land than our cities can support, and that we have an unrealistic standard of living which the environment cannot sustain and the rest of the world can never achieve'.[48] He suggests that the monoculture necessitates heavy pesticide treatment and that it must be abandoned. Many others see a return to rotation of crops and mixed farming as the only way out. The middle-of-the-road view is that a more sparing use of pesticides could avoid the evils including the appearance of resistant insects forms. The attempt to produce a bio-degradable pesticide has led to the use of organic phosphates and carbonates, but these are acutely toxic to man and animals and to useful insects. Recent research indicates that the pests could be eradicated without chemicals by using sex-attracting devices and releasing competitive radiation-sterilized males.[49]

Finally, mention must be made of defoliants. Like DDT these have been of great value, e.g. for the clearance of scrub and bushy weeds on tree plantations, but hundreds of reindeer have been killed by eating leaves which contained 25 ppm 2,4-D ad 10 ppm 2,4,5-T, nine months after spraying.[50] Dioxin persisting in vegetable oils prepared from plants which had absorbed some 245-T (in which dioxin is an unwanted contaminant) caused the death of millions of chicks in 1957.[51] It is clear that scientists have produced a very powerful weapon in these defoliants, but that they are far from understanding its full consequences. The use of defoliants in war, where one spraying kills all the trees in a mangrove forest,[52] constitutes a revolting misuse of a scientific discovery.

Radioactivity

The emission of radioactive material must be treated separately because the poisons may be passed via air, water, or the ground. The explosion of nuclear bombs in the atmosphere, whether for war or as a preparation for war or even as a 'scientific experiment', certainly distributes poisons which can have halflives of many hundreds of years and spread all over the world. Calder states that every young person who grew up anywhere in the world during the bomb testing period has a detectable amount of radio strontium replacing some calcium in his bones.[53] This is, of course, far below the medically significant level, but an accumulation of such effects will eventually undoubtedly have subtle psychological or genetic results which can only be harmful. It is almost certain that the inevitable consequence of releasing the existing stockpiles of nuclear weapons would be the complete destruction of human life within a few years, as well as the destruction of all mammals and birds.

The other threat to a human life of quality is the result of poisons produced by nuclear fission power stations. It is true that the danger has been recognized since the beginning and that more stringent standards have been applied to nuclear energy than to any other system. On the other hand, it has been decided by governments that we must have nuclear energy as our main source of energy for the future, and vast sums have been spent on research on this subject, a small fraction of which could have solved the problems of converting coal into electricity without men working underground or emitting SO_2 into the atmosphere. Moreover, this has meant that the committees deciding on the safety standards tend to be the officials of the bodies whose objective is to develop nuclear power rather than disinterested guardians of the public.[54]

The main problems are:

1. Radioactive contamination of the environment due to the fact that no shielding is perfect and that minor escapes of slightly radioactive material can occur. Here the US National Committee on Radiation Protection has established maximum permissible concentrations (MPC) which are at a level which cannot be proved to have a detectable biological significance. On the other hand, Tamplin of the Lawrence radiation laboratory maintains that pollution of a hundredth of the MPC may result in damage to humans because food chains are biological concentration mechanisms.[55] Loevinger points out that the breeder reactor will not only be necessary to conserve U^{235} but also will probably be cleaner than the conventional reactors, and yet only one breeder reactor has been included in a hundred applications to the AEC for authority to construct commercial nuclear power plants.[56]

2. The possibility of a serious accident to a nuclear reactor. In spite of all precautions serious accidents *can* happen, especially in a new technology. Sabotage or shelling of a nuclear reactor in a coup d'état, furthermore, could lead to a very serious escape of radioactive material.

3. Disposal of radioactive waste products. The European Nuclear Energy Agency has dumped many tens of thousands of sealed oil drums containing altogether a few tens of thousands of curies low-activity waste. These drums will certainly disintegrate in a few years. The high-activity waste is the direct product of the nuclear fission process. It is a very serious problem since the radioactivity can heat the containers to 315–500 C. Significant radioactivity is retained for 500,000 years.[57] It is estimated that the present US civilian commercial plants will produce 58,000 cubic feet per day by AD 2000. There is certainly no known

way of ensuring that this perpetual furnace can be contained for more than a century or two. We are thus leaving an unsolved problem for future generations. The AEC plan to store the hot material in canisters set into holes drilled in the floor of a salt mine in Kansas and to fill the vault with salt. The British store fission-contaminated fuel casings under water in concrete silos and the solution in water of unwanted elements from the spent fuel in stainless-steel lined concrete tanks with internal coolers. The rate of increase of the stored material is 45 m^3 per year. The total radioactivity stored exceeds 10^8 curies, which can be compared to the estimated total radioactivity of all the world's oceans, or 10^{11} curies. Thus, according to Windscale, already 0.1 per cent of the total radioactivity of the world's oceans is stored within a few hundred cubic metres. These tanks will have to be tended for 500 to 1000 years.[58] Glassification of the waste into solid ceramic bricks may provide a less hazardous solution of the problem, but even that precaution has not yet been achieved.

Conclusions and recommendations

1. There is little doubt that the most serious problem for humanity that arises from our thoughtless pollution of the environment is not the occasional spectacular case where a few hundred people are poisoned by some gross negligence. It is the slow but steady attrition of the quality of life of all humans, especially those who have to live or work in towns and crowded areas. Among the worst contributors to this deterioration (which occurs for rich and poor alike) are noise, SO_2, Pb, CO, and other products of incomplete combustion of hydrocarbons and the deterioration of food quality caused largely by the excessive use of pesticides and bactericides and the economic pressure to have 'factory farming.'

2. Whereas science is an international profession, there must be a professional ethic for scientists to serve, to the best of their ability, the needs of *all* human beings of the present and the future. The immediate problem, to which all scientists who accept this ethic must give their full attention, is how to halt and, indeed, reverse this deterioration in the quality of life in spite of the trend for the world population to double before a better standard of living and better education can spread sufficiently to cause any deflection of this trend.

3. Whereas scientists cannot claim to be experts in economics or politics, they can, and must, make quite clear to the general public what technological choices lie before us. On the one hand we can continue to demand still further possessions for those who have good jobs in technology-based societies at the expense of increasing pollution, unemploy-

ment, and poverty of the societies that have not many machines, with the probable result of the premature exhaustion of raw materials and unlimited arms escalation. It is quite certain, from what we learn from science and the study of our technologies, that none of these problems can be solved by governments making laws because these problems are the inevitable consequences of the basic idea of both capitalist and socialist systems that the success of an individual is to be assessed by his accumulation of personal possessions. Prohibition and anti-drug laws are clear examples of the inadequacy of legal action to deal with a problem without the full backing of public opinion. Pollution is just the same sort of problem.

4. There is, in fact, a clear and simple choice before every individual exemplified by the motor car. One can choose the high-acceleration status symbol which is replaced every two or three years; or one may demand a small, safe, quiet, non-polluting, reliable car built to last a lifetime. Until the vast majority of people, who are in a position to choose, choose the good-neighbour car, we shall not avoid pollution by laws alone.

5. The only hope for the survival of humanity with a life of reasonable quality is the idea of spaceship earth – the devotion of enough scientific effort (together, of course, with the full efforts of all other professions) to achieve a permanent equilibrium between man and his environment. This means that success in life must be judged primarily by the contribution each individual makes towards achieving this equilibrium.

6. Instead of exerting pressure on industry or the farmer to produce as much as possible at as low a cost as possible, we have to produce a climate of public opinion in which the success of the producer is judged by:

   a. The quality and safety of the product.
   b. The avoidance of all bad by-products such as pollution, noise, and accidents.
   c. The minimum use of limited raw materials, such as minerals, oxygen, land, water, fuel.

7. The particular responsibility of the scientist and engineer is to combine the following branches of knowledge:

   a. The invention of the machines necessary to make what we need without noise, danger, pollution, and ugliness.
   b. A complete understanding of worldwide ecology.
   c. An equilibrium fuel system, i.e. one based essentially on solar energy.
   d. The recycling of all metals and other materials in short supply.

e. The adoption of agricultural methods that can provide food of the finest quality indefinitely from the same soil.

f. The re-use of all organic refuse.

Only in this way can we find a feasible means of enabling seven billion people to live lives of plenty and good quality on the limited surface of the earth in permanent equilibrium with the environment.

8. Scientists must use their entire persuasive force to get governments to allocate much more money to research on the problem of man's living in equilibrium with the environment. This is a far more urgent problem than any in the fields of economics, weapons, or scientific curiosity about the working of the universe. Scientists can no longer allow themselves to choose their problems only by their intellectual excitement; they have to mobilize all their skill to save humanity from self-destruction. It is nonsense to suggest that scientists cannot make a sufficient contribution to this problem. When one looks at the immense change in our way of life brought about by science and technology in the last two centuries, it is perfectly clear how great their contribution is and must continue to be.

CITED REFERENCES

1. W. S. Broecker, 'Man's Oxygen Reserves,' *Science* 168, June 1970, p. 1537.
2. L. Machta and E. Hughes, 'Atmospheric Oxygen in 1967—1970,' *Science* 168, p. 1582.
3. L. Machta and E. Hughes, *Futures* 2, No. 2, June 1970, p. 106.
4. *Idem, Scientific American,* September 1970, p. 131.
5. *Idem, MIT Study,* SCEP.
6. *Ibid,* p. 91.
7. J. Simproner and W. L. Woodley, 'Seeding Cumulus in Florida,' *Science* 172, April 9, 1971, p. 117.
8. L. B. Lave and E. P. Seskin, 'Air Pollution and Human Health,' *Science* 169, August 21, 1970, p. 723.
9. G. H. Bosanquet, 'The Flow of Chimney Gases,' *Air Pollution,* London, Butterworth Publishers, 1957, p. 114.
10. R. E. Newell, 'The Global Circulation of Atmospheric Pollutants,' *Scientific American,* January 1971, p. 32.
11. K. Westberg, N. Cohen, and K. W. Wilson, 'CO: Its Role in Photochemical Smog Formation,' *Science* 171, March 12, 1971, p. 1013.
12. Alice Garnett, *Air Pollution,* London, Butterworth Publishers, 1957, p. 97.
13. *Ibid.,* p. 97.
14. R. E. Newell, *op. cit.,* p. 32.
15. R. E. Inman, R. B. Ingersoll, and E. A. Levy, 'Carbon Monoxide Sink,' *Scientific American,* August 1971, p. 47.

16. H. Reiquaim, 'Sulphur: Simulated Long-range Transport in the Atmosphere,' *Science,* 170, October 16, 1970, p. 318.
17. Daniel Cordtz, 'Autos: A hazardous Stretch Ahead,' *Fortune,* April 1971, p. 68.
18. P. T. Sherwood and P. H. Bowers, 'Air Pollution from Road Traffic,' Road Research Labs Report, LR 352, Crowthorne, Berks., p. 25.
19. *Anon.* 'Automobiles et nuisances: pour un program d'action,' Paris, *Mission interministerielle pour l'environment,* 1971, p. 5.
20. Murozimi et al, *Geochim. cosmochim. Acta* 33, 1967, p. 1237.
21. D. Bryce Smith, 'Land Pollution: A Growing Hazard to Health,' *Chemistry in Britain* 7, No. 2, February 1971, p. 54.
22. J. Dunne, 'The Stratified Charge Engine,' *Popular Science Monthly,* May 1970, p. 55.
23. Daniel Cordtz, *op.cit.,* p. 68.
24. M. E. S. Fayed, M. A. Salen and L. Otter, 'A Study of Air Pollution Around a Sulphire Acid Plant.' *J. Inst. F.* 44, March 1971, p. 143.
25. H. Reiquaim, *op.cit.,* p. 319.
26. A. J. Robinson, 'Air Pollution,' *Journal R.S.A.* 119, July 1971, p. 513.
27. R. E. Pattle and H. Cullumbine, 'Toxity of Atmospheric Pollutants,' *British Medical Journal* 2, 1956, p. 913.
28. J. Pemburton and C. Goldberg, *British Medical Journal* 2, 1954, p. 567.
29. A. M. Squires, 'Clean Power from Coal,' *Science* 169, August 28, 1970, p. 821.
30. M. K. Hubert, 'Energy Resources,' Chapter 8, in *Resources and Man,* San Francisco, W. H. Freeman, 1969, p. 159.
31. *U.S. Water Resources Council Publication,* 1968, p. 4.
32. R. W. Holcomb, 'Waste Water Treatment,' *Science* 169, July 31, 1970, p. 459.
33. J. H. Ryther and . M. Dunstan, 'Eutrophication Reconsidered,' *Science* 171, March 1971, p. 1008.
34. E. E. Angino et al., 'Arsenic in Detergents,' *Science* 168, April 17, 1970, p. 389.
35. J. A. Tolley and C. D. Reed, 'Lead Pollution at Crisis Point,' *Journal Royal College of General Practitioners,* May 21, 1971, quoted in *The Sunday Times* (London), April 25, 1971, p. 13.
36. A. Curley et al., *Science* 170, April 1971, p. 65.
37. L. J. Goldwater, 'Mercury in the Environment,' *Scientific American,* 224, May 1971, p. 15.
38. F. J. McClure, *Water Fluoridation,* Superintendent of Documents, Washington, D.C., p. 304.
39. Hodge and Smith, *Fluorine Chemistry IV,* p. 452.
40. Pinet, *Fluoride Quart. Report* 2, 1968, p. 86.
41. H. A. Cook, *Your Environment — Fluoride Toxicity,* London, 1970, p. 78.
42. P. H. Abelson, 'Marine Pollution,' Editorial, *Science,* 171, January 8, 1971.
43. P. Beaconsfield, 'Internal Pollution: Our First Priority,' *The New Scientist* 171, 8 March 1971, p. 600.
44. Metcalf and Pitts, *Advance in Environmental Sciences* 1, p. 21.
45. S. G. Bloom and D. B. Menzel, 'Decay Time of DDT,' *Science* 172, April 16, 1971, p. 213.
46. *Report of the Royal Commission on Environmental Pollution,* H. M. Stationery Office, London, February 1971.
47. F. F. Darling, 'Land Pollution,' *JRSA* 119, July 1971, p. 520.
48. J. C. Dickinson, 'Hydra-Headed Pesticides,' *Science* 171, January 8, 1971, p.
49. P. H. Abelson, 'Control of Agricultural Pests,' *Science* 171, February 5, 1971, p. 437.

50. D. Jones, 'The Defoliant Story: A Cautionary Tale,' *Your Environment,* 1970, p. 118.
51. H. L. Harrison. 'Systems Study of DDT Transport,' *Science* 170, October 30, 1970.
52. P. M. Bofley, 'Herbicides in Vietnam,' *Science* 170, January 8, 1971, p. 43.
53. Ritchie Calder, 'President's Address,' *Conservation Society,* November 23, 1968, p. 15.
54. P. M. Boffey, 'Radiation Standards: Are the Right People Making Decisions,' *Science* 171, 1971, p. 780.
55. A. R. Tamplin, 'They Won't Shut Up,' *New Scientist,* May 27, 1971, p. 532.
56. L. Loevinger, 'Nuclear Power and the Public,' *Science* 171, February 26, 1971, p. 795.
57. C. Holden, 'Nuclear Waste,' *Science* 171, April 16, 1971, p. 249.
58. W. C. Patterson, 'Hazards of Radioactive Waste,' *Your Environment,* London, 1971, p. 99.
59. 'Air Pollution from Road Vehicles,' *Report by Technical Committee of National Society for Clean Air,* 134 North Street, Brighton; 'The Automobile and Air Pollution,' 'U.S. Dept of Commerce, Washington D.C., U.S. Govt. Printing Office, October 1967,' Chapter VII, *The Federal R. and D. Plan for Air Pollution Control, Reciprocating Internal Combustion Engines,* Battelle Inst., January 11, 1971; R. E. Newell, 'The Global Circulation of Atmospheric Pollutants,' *Scientific American,* January 1971; D. Cordtz, 'Auto — A Hazardous Stretch Ahead,' *Fortune,* April 1971, p. 68; A. J. Haagen Smit, 'Air Conservation,' *Scientia* CIII, 1968 ,p. 261; B. W. Duck and C. L. Bailey, 'An Oil Company's Viewpoint,' *2nd Internat. Road Traffic Conference,* Italy, October 1967; B. L. H. Bishop, 'Control of Crankcase Emissions,' *Inst. Mech. Eng. Proc.* 1968—69, pt. 3E, p. 183.

W. LENZ

Chemicals as a Cause of Human Malformations

If thou couldst, doctor, cast
The water of my land, find her disease,
And purge it to a sound and pristine health,
I would applaud thee to the very echo,
That could applaud again.
(Shakespeare, *Macbeth*, V, iii, 50)

The idea that chemical agents might severely damage mental and physical development in man is by no means new. Morel, the French psychiatrist, in his *Traitê des dégénerescences physique, intellectuelle, et morale de l'espèce humaine,* published in 1857, collected much material in an attempt to show that alcohol, hashish, lead, and other poisons may produce degeneration, i.e. congenital abnormalities of development. Ballantine in his *Manual of Antenatal Pathology and Hygiene,* published in 1902, mentioned lead, mercury, phosphorus, arsenic, copper, sulfuric acid, carbon monoxide, chloroform, ether, morphium, quinine, sodium salicylate, cocaine, and other drugs as agents potentially harmful to the human fetus. He discussed such effects, concluding: 'It is, therefore, most unsafe to attempt to form any general conclusions regarding the effects of poisons on the unborn infant. All that may with any assurance be said is that there is experimental proof that some poisons reach the fetus and that sometimes these poisons produce structural alterations in the fetus and placenta, and that clinical evidence to a certain extent justifies us in asserting that a similar transmission and similar effects may be met with in the human subject.'

Since 1902, much more knowledge about teratogenic actions of chemicals has been accumulated as a result of many experiments performed in laboratories throughout the world. Thus, in 1950, Ancel in his monograph, *La Chimiotératogenèse, réalisation des monstruosités par des substances chimiques chez les vertébrés,* analyzed the results of experiments reported in 490 published papers. In 1958, Professor of Pathology Willis, in *The Borderline of Embryology and Pathology,* pertinently remarked: 'It will be noted that only a few chemical agents have yet been tested for teratogenic effects in mammals, and that these do not include any of the commonly used alkaloids, sedative drugs, sulphonamides, antiseptics, dyes, organic solvents, or metallic and other inorganic sub-

89

stances. Methodical investigation of the effects of all of these on early mammalian development is needed.'

This was written at about the time when thalidomide was first put on the market, and Professor Willis's remark certainly was not based on knowledge that was not available before 1957. Of fifteen papers quoted in Willis's chapter on 'Chemical Poisons as Causes of Malformations,' fourteen had been published before 1955, and only one in 1957. Willis was not a single voice shouting in the desert. Many scientists very clearly expressed similar opinions in the fifties, and these opinions have been echoed in many medical journals.

In the apologetic literature published after the thalidomide catastrophe, it has sometimes been stated that, at the time when thalidomide was introduced, nobody thought, or could have thought, of ill effects of drugs on the embryo. Such statements are obviously untrue. The truth is that far less attention has been paid to the potential dangers than to the potential usefulness of drugs. Both desirable and undesirable effects of chemicals were, as a rule, unpredictable from the knowledge of the chemical formula alone. In thalidomide, the desired effect, i.e. its sleep-inducing action, seems to have been found as a side effect that had not been initially looked for. In fact, there was no scientific evidence from animal experiments to show that thalidomide might induce sleep. Before thalidomide was put on the market, neither its potential usefulness nor its potential danger had been studied by scientific methods.

Why has more attention not been paid to the vast literature on experimental teratology? The answer to this question is apparently not to be found in science or logic, but in extra-scientific interests. As this is a question of value and a legal problem rather than one of facts I will not, in this context, discuss this point further.

Few chemicals are definitely known to be teratogenic in man. These are aminopterin, busulfan, androgenic hormones, and thalidomide. The total number of cases of malformations attributed to aminopterin and busulfan is less than ten, the number attributed to androgenic hormones less than two hundred. Androgenic hormones affect only the external genitalia and behavior, usually in a minor degree without serious consequences. We might thus confine the discussion to the thalidomide tragedy, which hit certainly more than 5000, but probably less than 10,000 children in many countries. The thalidomide experience has had a profound effect upon the attitude of pharmaceutical firms, health authorities, physicians, and the public. It has served to make everybody more cautious, to demand more knowledge of side effects and better controls of drug trade. The state of affairs, however, is not yet entirely satisfactory. Most experts feel that we have insufficient knowledge to make the occurrence of another, similar event impossible.

Often, however, the tacit implication is that progress in developing ever more new drugs must by no means be hindered even though it might involve the risk of another tragedy. My presentation is meant to show one side of the balance which has to be weighed against the usefulness of drug therapy.

The first victim of thalidomide, in 1956, was a girl born in Stolberg, (the place of the German thalidomide producer). It happened almost one year before thalidomide was put on the market, because the child's father had been working for the thalidomide-producing firm for several years and had been given samples (thalidomide was then still in the experimental stage) by one of the company's physicians. The first clinical and experimental papers on thalidomide appeared several months after the mother had taken the drug. The daughter has almost complete absence of the auricles, atresia of the external meatus, and malformation of the ossicles with hearing loss due to defective conduction (absence of the external ear, closure or nondevelopment of the ear, and malformation of the middle ear). These associated defects have been shown to be typical of thalidomide ingestion on the 35th postmenstrual day.

The next case of which I know is a girl born in 1958. She has bilateral aplasia (nondevelopment of the shorter and thicker of the two bones of both forearms) of radius and thumbs, dislocation of the shoulders, and club hands; this is the type of arm deformity best known to be associated with thalidomide intake. The mother's physician, who was married to a representative of the German thalidomide producer, had obtained experimental samples of thalidomide and given it to the mother on September 9, 1957 — three weeks before the drug was put on the market. This was the mother's 42nd postmenstrual day, almost exactly the same date as would be inferred from the type of malformation as the most probable date of its causation.

In the following years, between 5000 and 7000 thalidomide children were born in West Germany. When sales figures of thalidomide became available, a comparison could be made between the monthly sales figures and the birthdates of these children. The two curves are strikingly similar in several characteristics: both show an initially slight, and subsequently steeper rise, a sharp peak, an initially slight and subsequently steeper fall, and an abrupt end. For all of these six characteristics the curve of thalidomide sales was followed about nine months later by that of the birth of thalidomide children.

Figures of thalidomide sales have not been made officially available. I obtained these figures, however, through a leakage in the system of top secrecy which usually veils the sales figures of drugs almost as effectively as those of explosives for military purposes. The predicted end of the epidemic was on July 27, 1962, i.e., 280 minus 35 days after November 25, 1961, the day when thalidomide was withdrawn from sale in West Germany. The average interval between the first day of the last menstrual period and the date of birth is 280 days; 35 days is the mean interval between the beginning of the last menstrual period and the beginning of the sensitive period of the embryo, when only ears, eyes, and facial muscles are affected. As was to be expected, the epidemic abruptly ended by the end of July 1962, and, significantly, in the last children born with thalidomide embryopathy usually only the ears were affected.

A complete end of the epidemic, however, could have been expected only if the warning had reached everybody. This was not the case. The health authorities, the doctors, and the manufacturing firm were rather reluctant to give a thoroughly effective warning immediately. As a result, several expectant mothers had not heard of the teratogenic activity of thalidomide and continued to take the drug. Thus their children were born with typical malformations after the end of the sensitive period. The last child with thalidomide deformities was born in Germany on February 29, 1964. In this case, the mother, in early pregnancy, took, for suicidal purposes, the rest of a bottle of thalidomide emulsion which had been prescribed to her in 1961. Her child was born with triphalangeal thumbs (three-sectional, that is, formed like the other fingers).

The country with the highest number of thalidomide-deformed children, besides Germany, is Japan. The special problem in Japan is that thalidomide continued to be sold for about one year after it had been withdrawn in Germany, Sweden, and England. The peak incidence of Japanese thalidomide malformations apparently occurred in the latter half of 1962, i.e. at a time when virtually no more cases were born in Europe. A discussion of the situation in Japan, however, does not appear to be appropriate at the present time, because litigation against Dainippon and the Japanese Government has been going on at the Tokyo District Court since February 1971. Moreover, the Düsseldorf County Court has decided to threaten me with either imprisonment for up to half a year or an unlimited fine, if I should repeat my previously stated opinion of Dainippon's behavior. I have, so far, not received from the Düsseldorf County Court an understandable explanation of why my opinion should be wrong, or why stating it should be punishable. I cannot, therefore, offer juridically correct explanations of why it was in order for Dainippon to continue to sell thalidomide in view of its potential danger and why criticism thereof should be an offense to be punished. I hope that this matter will be clarified in the future.

I have at hand detailed information of malformed children born in Japan after July 1962, because Dainippon decided to leave thalidomide on the market. These include: Hidenori N., born on September 17,

1962 (whose mother took thalidomide between the 47th and 49th postmenstrual day); Yumiko K., born on September 18, 1962, and Akimi N., born on October 14, 1962 (intake of Isomin beginning on 47th postmenstrual day); Emiko F., born on October 18, 1962, and Maumi, S., born on November 25, 1962 (prescription on 44th postmenstrual day); Kenji, I., born on February 14, 1963; and Ryoko, T., born on April 16, 1963.

All these, and many more Japanese thalidomide cases, are exactly comparable in every detail to the thalidomide cases observed in other countries. The malformations following thalidomide ingestion by the mother are closely similar in all races. This is illustrated by cases such as the following:

Austria: Martin W., born on September 16, 1960, is a boy with defective upper and lower arms and absent thumbs. The physician had prescribed thalidomide to this boy's mother and to another woman who also had a baby with malformed limbs.

England: Graham, H., born on June 29, 1962, whom I have seen in Southampton, has severe defects of both arms and legs. His mother took thalidomide from the 43rd to the 49th postmenstrual day.

A second case from England, born on June 5, 1960, is patient no. 10, of Dr. Speirs, who had seen eighteen similar cases and found evidence of thalidomide intake in all of these except one (which was atypical in so far as it was unilateral). In the case shown, a pillbox containing thalidomide has been found by Dr. Speirs which was dated on the 50th postmenstrual day. Both upper and lower arms and the right thigh of the child had bone defects.

Ireland: Cliona R., born on February 24, 1961, without thumbs and with club hands. Her father is an Irish physician who had received thalidomide samples from a representative of the firm.

The Netherlands: Cornelia van R. was born on February 2, 1962, with severe defects of arms and fingers, malformed hip joints and thighs. Her mother took thalidomide.

Norway: The mother of this girl, born on April 5, 1962, took neurodyn tablets during the first weeks of pregnancy. Thereafter, she read a newspaper report about malformations following the intake of thalidomide. She was frightened and brought the article with her when she was admitted to hospital for delivery. The child's arms were rudimentary, the tibiae (the larger of the two bones of the lower leg) were lacking, there was hemangioma (vascular tumor) on her upper lips.

Such cases with a history of thalidomide intake given before birth and the baby showing the typical malformations are, of course, one of the strongest arguments in favor of a causative connection. Many such cases have been observed in Germany. I know of two such cases in Hamburg in which the obstetrician, motivated by the mother's history of thalidomide intake, was able to make a prenatal X-ray diagnosis of bone defects of the fetus.

Portugal: José Jorge M., born on October 13, 1961, has no thumbs and severely malformed bones of the forearms. Thalidomide had been prescribed to his mother in early pregnancy.

Iran: Ali Reza T. was born in Iran. His mother was in Hamburg, Germany, during the second month of her pregnancy. A photocopy of the thalidomide prescription is available. There is a malformation of thumbs and forearms with associated defects of eye muscles, hemangioma, and lack of tears, as often seen in other thalidomide children.

Pakistan: Wolfgang R. was born on August 12, 1962. His parents were living in Pakistan. At a time when thalidomide had just been banned from the German market, his mother received the drug from a Pakistani doctor. She took it from the 42nd to the 47th postmenstrual day. The child's arms and thumbs were underdeveloped.

We should keep the individual cases in mind in order to be able to grasp the human meaning of the total numbers of thalidomide babies from countries for which thalidomide sales figures are available. Both are roughly parallel. A closer correspondence can scarcely be expected, because reporting of cases was far from complete in most countries, and the attitude of pregnant women toward drugs differs in different countries. Thus, few thalidomide babies were born in London in spite of a rather high thalidomide consumption by the London population. Professor Penrose has kindly explained to me that in London's obstetric teaching hospitals it has been the practice not to administer any drug to pregnant women unless it is unavoidable. Pregnant women in Switzerland and in Austria have, according to doctors from these countries, been similarly reluctant to use drugs. The biggest part of the total thalidomide consumption has been by old people, especially in mental hospitals. If we could have a reliable estimate of the minor part taken by women of reproductive age, we would be in a better position to establish a biologically meaningful correlation between malformations and thalidomide consumption.

Yet, the overall figures are not devoid of value, particularly if consecutive years in the same country can be compared. In Sweden and the Netherlands there was definitely less increase in thalidomide consumption between 1959 and 1961 than in Germany. As a result the total number of thalidomide babies born in the years 1960, 1961, and 1962 has remained much the same. In England, both curves were steeper. In Canada, thalidomide was introduced in 1961. Thus, the first Canadian thalidomide children were not born before November

1961. As the Canadian government was rather slow in asking the producer to withdraw the product, 29 Canadian thalidomide children were born after July 1962. Similarly, thalidomide was only removed from the Italian market in July 1962, so that in Italy thalidomide children continued to be born up to the first months of 1963. The situation was similar to that in Japan. In both countries more than one firm was involved. Fourteen Japanese and ten Italian firms produced or received permission to produce thalidomide.

Figures from countries in which more than one hundred thalidomide children were born, such as West Germany, Japan, Britain, Sweden, Canada, and possibly Italy, lend themselves more easily to statistical arguments. In countries in which thalidomide was not a great commercial success, such as Denmark, Finland, Norway, Switzerland, Portugal, Mexico, the USA, Israel, and Lebanon, nevertheless, a few cases have been found associated with well-documented evidence of thalidomide intake.

In 1962, when a few people still honestly doubted the teratogenic effect of thalidomide, such single cases have been particularly convincing. The association of special limb deformities with thalidomide intake in Norway, Switzerland, and Portugal is sufficient proof in itself of the specific hypothesis formed previously on the evidence of similar observations in Germany.

The value of a hypothesis lies in its predictive power. On the basis of the experience in Germany, one could predict that similar experiences would be had in other countries in which thalidomide was used. This was found to be so. The critical statistical question, then, is whether such an association could also be expected on a null hypothesis, i.e. if no causative connection existed between thalidomide intake and malformations. The cases in Norway, Switzerland, and Portugal certainly cannot be reconciled with a null hypothesis. Nor is there any hypothesis, except the etiological one, which does make sense.

I should like to pay tribute to the newspapers who have greatly contributed to speedily removing thalidomide from the market in several countries and who have given the most efficient warning to expectant mothers. One wonders when Dainippon would finally have decided to withdraw Isomin if it had not been for the repeated attacks by the Asahi Shimbun Press. At that time Japanese doctors and health authorities kept remarkably quiet with the laudable exception of one young Japanese pediatrician, Dr. Tadashi Kajii of Sapporo. The German producer has confessed in a letter to the Japanese licensee that it was forced by the press campaign to withdraw thalidomide, though this was at variance with the more innocent version offered to the German public. Rather unexpectedly, from the scientist's point of view, the producers have been neither very efficient in collecting information on the teratogenic side effects of their products nor very critical in evaluating the facts. It is difficult to know whether lack of information, imperfect understanding of the facts, wishful thinking with emotions overruling logic, or other motives can explain the striking divergence between papers or statements published by the producers and those of independent scientists who worked on the problem. One more general lesson stands out: the decision to remove a potentially dangerous drug from the market must not be left to the producers; they may be too biased.

I do not believe that the scientist should be a moralist throwing stones at other people. His task is to understand rather than to condemn, to convince rather than to punish. If, however, after critical examination of the problem, the scientist is convinced that certain actions or omissions have practical consequences which are unacceptable in any system of human values, then his duty is to ask precise questions, to insist on precise answers, to find out the facts, and to help the public to know the whole truth and nothing but the truth. This is his specific contribution to social responsibility.

RUTH MULVEY HARMER

Selling Death

'Pest control, as it exists today, is largely a matter of merchandising.' I was a little startled and more than a little skeptical several years ago when I heard that assertion made by Dr. Robert van den Bosch in a California courtroom.[1] Since then, however, I have been convinced of the accuracy of that appraisal. Even more, as a result of the brilliant sales campaign carried out since World War II by the chemical industry, the planet has been inundated with a tide of poisons so widespread that there is no escape and so dangerous as to threaten every living thing and even life to come.

No rational person in the twentieth century would do away with chemical pesticides any more than he would eliminate the chemotherapeutics that have revolutionized the practice of medicine. No rational person should defend their misuse and overuse since, like 'wonder drugs,' agricultural chemicals have extraordinary power to harm as well as heal.

Yet the sales campaign has made an article of faith of the fallacious notion that without the unrestricted use of chemical pesticides the world would succumb to the apocalyptic disasters of starvation and deadly diseases. Although a wide variety of safe, economical, and effective measures does exist to curb the pests that threaten human health and food supply, the fallacy is so entrenched that to oppose the indiscriminate use of chemical poisons is to be branded 'unscientific' at best and 'unpatriotic' at worst — derided as a 'kook' or a 'Commie'. The extent to which that sort of thing occurs in the United States was made clear in the spring of 1971, when Senator Edmund Muskie revealed that the Federal Bureau of Investigation had prepared elaborate 'intelligence reports' covering the organizers of Earth Day observances in 1970. They linked 150 Senators and Congressmen and the hundreds of thousands of loyal citizens who protest pollution with extremist and subversive organizations and activities. Aside from being 'a monumental waste of taxpayers' money,' Senator Muskie said, the FBI performance was 'a dangerous threat to fundamental constitutional rights.'[2]

To question the activities of pesticide pushers is dangerous; to applaud them, a guarantee of 'right thinking' and often 'instant fame' — however preposterous the argument used.

Dr. Norman E. Borlaug, the 1970 Nobel Peace Prize winner for his contribution to the development of high-yield wheat strains, won front-page headlines all over the world in the fall of 1971 with his attacks on 'environmental nuts' and 'so-called ecologists,' who were out to eliminate DDT and the other pesticides. At a news conference arranged by Montrose Chemical Company, a major DDT manufacturer, and later at the biennial conference of the United Nations Food and Agriculture Organization, Dr. Borlaug held DDT chiefly responsible for staving off 'world chaos' and starvation for a quarter of a century and saving more than a billion people from malaria. 'There is no evidence in man that DDT is causing cancer or genetic damage,' he said.[3] 'There has not been one shred of reliable evidence that DDT has put any species of wildlife in danger . . . the argument that pesticides are upsetting the balance of nature is utter nonsense.'[4]

Similarly, two relatively unknown California State College science teachers achieved international attention a few months later with their remarkable argument that the thin shells of the eggs of the disappearing pelicans along the Pacific coast have been caused not by DDT, but by environmentalists and researchers who were frightening the birds![5]

The new media, often unduly sensitive to the demands of advertising and circulation departments, are less likely to give equal time for rebuttal. Yet it would be instructive to examine the arguments that are advanced by the chemical companies and by their loyal supporters in science, government, regulatory agencies, agribusiness, and even those ordinary citizens afflicted with a 'pesticide mentality.' Is the indiscriminate use of pesticides an economic imperative? Does the indiscriminate use of pesticides have no seriously adverse effect on human health? Is the balance of nature undisturbed by the indiscriminate use of pesticides?

When organic synthetic pesticides were introduced to the civilian market at the end of World War II, it was under the general title of 'economic poisons.' That identification helped to dissociate them from more traditional pesticides like arsenic and cyanide, which were generally regarded by lay persons as 'people poisons.' Moreover, it strengthened the chance of acceptance: after all, what is good business is good; what is economically feasible is desirable.

Farmers, notoriously resistant to new ideas, were targets of a massive and high-powered sales campaign. Pesticide advertising revenues transformed farm journals, the poor relations of publishing, into slick, glossy periodicals, whose advertising and editorial pages persuasively urged readers to buy and buy. In 1946, farm publications were assuring subscribers that DDT was the answer to all prayers. It was bringing aphids under control in Maine; eliminating flies and other livestock pests in Florida, Kansas, Texas, Arizona; wiping out corn borers in Iowa, tobacco pests in North Carolina, fruit moths in California and Georgia. If a little DDT was good, the farmers were persuaded, a whole lot more was better — and never mind the cost.

When the first glow began to fade, chemical companies had new answers. Only two years after agricultural bulletins put out by E. I. du Pont de Nemours & Company, Inc., were hailing the effectiveness of DDT for just about everything, they were touting with equal enthusiasm 'Marlate' for crops that were 'intolerant of DDT.'[6] Other pesticide companies followed the same track. Here was the perfect open-ended market. Chemicals that cost as little as 1.6 cents to produce could be sold to farmers for 51 cents or even more.[7] And as pests grew more resistant to cheaper pesticides, farmers had to use them in ever increasing amounts or turn to newer, more poisonous, and more expensive substitutes as the only way to insure bigger profits. Long-term costs could be reckoned later!

But later came sooner than farmers had thought.

As broad-spectrum pesticides like DDT killed off and crippled natural pest controls, U.S. farmers found the cost of protecting crops with chemicals more burdensome. (More sophisticated European farmers in countries like Switzerland, Italy, Holland, and Belgium rejected the 'cure-all-kill-all' substances, preferring narrow-range killers to help them.)[8] During the past few years the 'flight from the farm' in the U.S. has become a cause for national concern, with crop-protection as well as machinery costs becoming insupportable for farmers, chiefly small ones, but large ones as well. A striking example is offered by cotton farming since with that crop there has not even been the minimal control over pesticide application that has existed with foodstuffs that are checked for residues. What Dr. van den Bosch and others consider the 'landmark disaster' in cotton ecosystems occurred in the Cañete Valley in Peru, early in the 1950s.[9] Farmers there, who had been growing cotton for several decades with success, using such old-fashioned insecticides as calcium arsenate and nicotine sulphate, decided — were persuaded — at the end of World War II to shift to synthetic organics, principally DDT, benzene hexachloride (BHC), and toxaphene. First results were dazzling: pests died off rapidly. But success was short-lived, the old familiars developed resistance to ever increasing doses of poisons. Within a couple of years, BHC was ineffective against the cotton aphid; toxaphene was powerless to control leafworms. Not even the organophosphates to which farmers had turned after the organochlorines proved worthless could help. The yield for the 1955–56 season was the lowest recorded in the history of the valley.

Similar disasters have been occurring in other places: Texas, California, and in Central America, where countries like Guatemala and Mexico depend heavily on cotton for foreign exchange earnings. In northeastern Mexico, in the Matamoros-Reynosa and Tampico-Ciudad Mante areas, cotton has become an 'extinct crop.' The economic depressions that those areas are now experiencing as a result of what Dr. van den Bosch has called 'the insecticide treadmill,'[10] began when overuse of broad-spectrum pesticides converted the tobacco budworm, once a relatively innocuous pest, into a Number One enemy. As insecticides killed off its natural enemies, it increased in number; as the weaklings were eliminated, the tougher and more resistant strains were given license to explode. If Dr. Borlaug knows of no instance in which pesticides have upset the balance of nature, he simply hasn't taken a look around Mexico in recent years, although that is where he is headquartered.

But others have acknowledged that ecological backlash may transform 'economic poisons' into 'economic poison.' The Peruvian government was one of the first to halt the merchandising of chemical pesticides in an unrestricted way by encouraging the development of an integrated control program. This included a revision to the non-synthetic pesticides, improvements in agricultural practices such as burying and burning stalks to remove shelters for pests,[11] employing biological enemies of the pests to fight them. As a result, Cañete Valley cotton production reached high levels — the highest in history — and has since remained at peak.

There is no question at all, outside pesticide-pushing circles, about the desirability of integrated control: the use of biological, physical, and genetic means in combination with careful and limited use of chemicals to control pests. Simply pitting predators and parasites against agricultural pests, in California more than $100 million dollars was saved in less than ten years, according to the recent estimate of Dr. Paul De-Bach, University of California professor and president of the International Organization for Biological Control. A single species of parasite wasps he tracked down several decades ago, has saved California citrus growers an estimated $200 million.

Even those farmers who have been acutely afflicted with 'the pesticide mentality' have begun to call for integrated control as an answer to their problems. Among them are such wealthy and powerful men as Texas and California agribusiness leaders and Senator James Eastland of

Mississippi. He is one of the chemical industry's most influential friends in Washington and he has now joined forces with Senator Gaylord Nelson — perhaps the leading environmentalist in the U.S. Senate — to appropriate funds for research and development of integrated controls.

In some areas, unfortunately, conversion by economic suffering has come too late. A case in point is Central America, where the ecological backlash has created a public health menace of great magnitude. Recent studies indicate that the malaria increase in Central America has been associated with the increasing insecticide resistance of the disease-carrying mosquito *Anopheles albimanus*. In the summer of 1971, Dr. George Georghiou said in California, and later in Geneva to members of the World Health Organization, that the WHO anti-malaria campaign was in trouble in Central America because of the resistance of mosquitoes not only to DDT, but to new poisons — including organophosphates — used to control cotton pests. The widescale application for agricultural purposes, he said, has given the mosquitoes 'a kind of "basic training" and prepared them for the time when these poisons were directed against them.'[12] If similar resistance develops in the disease carriers in Asia and Africa, consequences would be catastrophic. Other scientists have expressed similar concern, pointing out that in many of the developing nations, the 'science' of pest control has taken such a merchandising bent that if problems arise the single solution is 'to call the biggest pest-control operator.'

Central America is not the only area rendered vulnerable to epidemics by the misuse of pesticides. Also in the summer of 1971, the moment that public health workers in California had feared for years finally came. The mosquito that carries western equine encephalomyelitis, a virulent sleeping sickness that attacks horses and humans had demonstrated its immunity to *all known chemical pesticides*.[13] State officials describe themselves as 'virtually at wits' end' and warned against the possibility of an epidemic in 1972 — one perhaps worse than the outbreak that occurred in 1953, when mosquitoes developed an immunity to DDT. At that time, 50 persons died and others were afflicted with the disease, which often causes severe mental damage.

It is highest irony that public health should be jeopardized by the overuse of pesticides since the principal reason for tolerating the poisonous assault has been the protection of public health. But it should come as no surprise at this late date. As early as 1946, less than two years after DDT was first used in a massive way by Allied troops to stave off a possible typhus epidemic in Naples, scientists from Sweden to Italy were issuing warnings about the resistance of disease carriers.[14] Very dramatic cases of backlash have been reported since then. One involved a malaria control program in Bolivia designed to kill rats. In

101

fact, it proved more successful in killing off cats, and, therefore, boomeranged. The 'aided' area was overrun by a mouselike mammal harboring black typhus virus, and more than 300 'beneficiaries' of the spray campaign died from that illness.[15] Delegates to the UNESCO commission's thirteenth national meeting in San Francisco in 1969 were told of another health campaign in which WHO workers went into some remote mountain villages in Borneo. Armed with DDT sprayers and impelled by a desire to rid the villages of flies and mosquitoes that carried sickness, they accomplished 'Mission Pesticide' in brief and heroic fashion. But something went wrong. The cats that ate the lizards that ate the poisoned roaches and flies began to sicken and die. And the rats that feared the cats, that were so important a natural enemy, were no longer afraid to go into the villages. They invaded them, bringing vectors of such serious ills that the threat did not subside until the Royal Air Force parachuted a new supply of cats into the village to restore a more equitable balance of nature.[16]

Pesticides, which are known in the trade as *economic poisons,* have a more direct impact on public health since — a point that is often minimized or ignored in discussing them — they are also *people poisons.* First reports of the adverse effects of DDT appeared in the *British Medical Journal* in 1945,[17] yet then and since research scientists have been scoffed at for suggesting that in addition to producing 'DDT jitters' the pesticide damages liver, heart, brain, and sex glands. A number of US scientists appeared at Congressional hearings instituted by Congressman James J. Delaney in 1950, 1951, and 1952 to testify that patients whose fatty tissue contained high levels of DDT showed signs of liver damage in addition to feelings of exhaustion, irritability, and mental dullness.[18] One expressed concern that DDT residues in foods might be responsible for the greater number of hepatitis cases.[19] Subsequent testimony was attacked with equal vigor by the chemical industry and allies in and out of government. Dr. Wilhelm C. Hueper, former director of the National Cancer Institute, told the Ribicoff Subcommittee in 1963 that his concern about the carcenogenic properties of DDT and other pesticides was so opposed by Washington policymakers that his research was halted, his staff of sixteen immobilized, and that he had even been excluded from official meetings because of his work.[20] In 1969, at hearings in Wisconsin on the Environmental Defense Fund petition for a ban on DDT, further effects were cited: Dr. Richard M. Welch, a pharmacologist at Burroughs-Wellcome Research Laboratories in New York, testified that DDT reduced the effectiveness of a number of drugs: and Dr. Robert Risebrough, University of California biologist, contended that sex hormones were affected by DDT.[21]

102

The chemical industry struck back through arguments advanced by its attorney that DDT critics were 'food faddists and/or health nuts, preoccupied with their own sexual potency.'[22] And it used as scientific basis for that kind of logic the fifteen-year-old study made by Dr. Wayland J. Hayes, Jr., 'proving' that 'a large safety factor is associated with DDT as it now occurs in the general diet.'[23]

The impact of the Hayes study cannot be ignored. His sampling was small and selection of the subjects questionable (it has been found that convicts participating in experiments are so eager to be cooperative that they fail to report adverse effects).[24] Some of the findings were highly subjective (Dr. Hayes dismissed the complaints of two of the subjects as 'obviously ... of psychoneurotic origin').[25] Yet the Hayes study has been the chief basis for the continuing argument that DDT is safe. That argument has had profound impact all over the world, since, as chairman of the World Health Organization's Select Committee on Pesticides, what Dr. Hayes recommended was what nations got.

From the standpoint of human health, DDT is one of the less acutely toxic of the pesticides now widely employed. Some of the other chlorinated hydrocarbons are far more poisonous — chronically and acutely. Almost thirty years ago, Dr. A. J. Lehman, then director of the Food and Drug Administration's pharmacology division, told the Delaney Committee that he 'would hesitate to eat food that had any chlordane residue at all.'[26] Yet Americans and other peoples have continued to do just that. They have also been widely exposed to lindane, which Dr. M. M. Hargraves of Mayo Clinic has implicated in hundreds of cases of serious blood disorders;[27] they have been exposed, sometimes fatally, to heptachlor, endrin, dieldrin, and aldrin. Those poisons have been so widely promoted as 'safe' that almost no laws now govern their shipment and storage. A major error! Some years ago, for example, when flour being carried by ship to Saudi Arabia and Qatar was contaminated by endrin stored nearby, 26 persons died and 874 were affected.[28]

Not even the organophosphate chemicals are treated with any realistic precautions, although some of the 'nerve gas' pesticides are so acutely toxic that a single drop on the skin is enough to kill an adult male. Because the effects of many of the organophosphate pesticides are so horrifying and because some of the symptoms are so obvious, it was assumed that the chemicals would be used under rigidly controlled circumstances and that victims could be readily identified and treated.

Neither assumption was warranted. Transported by air, rail, truck, and car and stored with incredible casualness, parathion is one of the major killers — and it will claim more lives as the chlorinated hydrocarbons are curtailed and eliminated. In one shipping accident some years ago, 106 persons died and 826 others became seriously ill, when a cargo of flour was contaminated by a nearby container of parathion.[29] More recently, a number of children were almost killed when several of them put on jeans contaminated with Phosdrin. They were part of a shipment stained eight months earlier when a can of the pesticide being carried in the same truck sprang a leak.[30]

The chemical industry and its apologists have protested all regulation, insisting that the 'small' number of victims does not warrant any. But the number of victims may be far higher than most have suspected. One expert, who testified before a Senate investigating committee in 1969, estimated that more than 100,000 cases of parathion poisoning occur each year — many of which are erroneously diagnosed as 'flu' by doctors who have had little training in the subject.[31] How many are actually affected is anyone's guess. Also in 1969, a doctor at a rural clinic in a California farm county said that nearly half of the children he tested during a nutrition study showed signs of organic phosphate poisoning: skin rashes, burning eyes, nausea, vomiting, dizziness, difficulty in breathing. 'Some of the children, like their parents, had been feeling sick,' Dr. Lee Mizrahi told legislators; few, however, would have sought medical care. He indicated his dismay and concern about the widespread poisoning: 'To me it is tragically absurd that in 1969 such a study by an obscure rural doctor should be the first one ever done on children.' He added: 'We think this problem is widespread.'[32]

Like the chlorinated hydrocarbons, the organophosphates and carbamates — another important category — have also been known to produce long-term effects. They have been described as carcinogenic; they have been associated with birth abnormalities in test animals. According to a recent report by British medical researchers, a group of persons exposed to parathion and similar pesticides for between 18 and 24 months suffered 'schizophrenic and depressive reactions, with severe impairment of memory and difficulty in concentration.'[33]

Mercury compounds, which have been used widely as fungicides, also affect vital organs, including the brain. Like many other pesticides, methylmercury is teratogenic, inducing birth defects when pregnant mothers are exposed to it.[34] So are the phenoxy herbicides. With all, the response of the pesticide industry has been the same. When it is no longer possible to pretend that they do not represent a threat to human health, the industry advances the curious argument that even though the evidence is damaging, no action should be taken 'until *all* the facts are in.' Translated, that means, of course, that no action should be taken — ever.

That the victims should defend the right of the Borgias of the en-

104

vironment to continue to poison them and that the victims are so willing to pay high prices for that non-privilege is a tribute to the persuasiveness of industry salesmen and to their allies in the all-out chemical war: the *agribusinessmen* who have transformed farming into the violent science in their pursuit of profits: the *scientists,* including educators, who have bartered objectivity and expertise for the financial rewards of pushing pesticides; the *regulatory officials* who have set aside their responsibility to protect the public and who have allowed the poisoners themselves to act as regulators; the *legislators* who have permitted industry lobbyists to draft pesticide laws and to set penalties for abuse — as a consequence of which it is easier to buy one of the deadly poisons than even a mildly potent medicine and it is cheaper to violate the laws than to obey them. Not finally, the allies include ordinary lay persons, who have been so brainwashed by the 'experts' that they have encouraged, subsidized, and actively participated in the extermination campaign against the planet and themselves.

The record of government agencies at all levels has been, for the most part, a dreary one in relation to the proper use of chemical pesticides. Even so reputable an organization as WHO has massively pushed poisons — partly as a result of it commendable goals and partly because of the special interests of some of its representatives. I was frankly startled by the response received from a WHO information officer in New York in the summer of 1969, when I called at the United Nations headquarters to find out what official policies were. Frowning at my questions, he brusquely dismissed 'people who worry about pesticides.' Pesticides, he concluded the discussion emphatically, are 'wonderful.' I was even more startled by some of the statements in a WHO report on the *Safe Use of Pesticides in Public Health.* Some of them — e.g., 'The safety record of DDT remains outstanding'[35] — echoed in an uncanny manner the 'briefing papers' put out by the National Agricultural Chemicals Association task force on that substance. But, as has been noted, it was less startling after I became aware that the chairman of the WHO expert committee was Dr. Hayes. (Not all UN agencies are enthusiastic about DDT. In the fall of 1969, Secretary General U Thant issued a report asking pointedly about the advisability of the Children's Fund annually distributing 12 million pounds of DDT to various countries. How long, the report demanded, should UNICEF continue the practice?)[36]

An awesome example of the pesticide pushing record of the US Department of Agriculture came to public attention a few years ago when Congressman L. H. Fountain of North Carolina had his subcommittee on government operations look into charges made by the General Accounting Office about the lax administration of the Federal

Insecticide, Fungicide, and Rodenticide Act (FIFRA) of 1947, as subsequently amended. GAO investigators reported that relations between pesticide makers and the USDA's Agricultural Research Service had been so amicable during the thirteen years from February 1966 through February 1968 that not a single violator had ever been reported to the Department of Justice for criminal prosecution. Yet major and deliberate violations were frequent. For example, during 1966, ARS tested 2751 of the 60,000 products then registered. Of that relatively small number, 750 samples were in violation of the law; 562, in major violation. Some of the offenders had violated the law up to twenty different times during the year.[37]

The focal point of the Congressional hearing was the Shell 'No-Pest Strip,' one of the most popular devices on the market — merchandised in the USA alone through 280,000 retail outlets. The golden cardboard rectangles dangling from the ceilings of millions of homes, restaurants, hospitals, and commercial establishments contain a strip of plastic impregnated with a highly toxic organic phosphate known as DDVT. (The company claims exclusive patent rights to that chemical although, according to a director of the US Public Health Service Communicable Disease Center in Georgia: 'It has been determined that the Federal Government is entitled to the entire right, title, and interest in the invention of DDVP.')[38]

According to the findings of the committee, in 1963 a pharmacologist in the Pesticides Regulation Division of the USDA urged that the strips carry a label with the word POISON and a picture of a skull and crossbones in order to indicate the high toxity of the product to humans. Although the US Public Health Service and health experts in other agencies agreed, the pharmacologist's protests were overruled by his superior — a man who soon after resigned from PRD and went to work for Shell Chemical Company. It was not until almost seven years later that the strips carried warning indications — a relatively mild note, cautioning against using them around food or in rooms where infants, ill, or aged persons are confined.

At the conclusion of the hearings, the Congressional investigators severely reprimanded the USDA Pesticides Regulation Division for having 'failed almost completely' to enforce the provisions of the act designed to protect the public from hazardous products and urged the Department of Justice to consider possible proceedings against a number of the principals in the affair — Dr. T. Roy Hansberry, Dr. Mitchell R. Zavon, and John S. Leary, Jr. All of them had been associated with Shell and with the US Department of Agricultural in clear violation of the 'conflict of interest' statute in the US Criminal Code which prohibits any officer or employee from participating in a proceeding, if he

106

has any financial interest in a particular matter — even rendering advice.

No action was taken — other than the transfer of the director of the Pesticides Regulation Division to another post for a brief period — until the storm blew over.

The record of the Department of Health and Welfare's Food and Drug Administration, the other principal agency involved in pesticide regulation for many years, has been equally dreary. 'What the FDA is doing and what the public think it's doing are as different as night from day,' a former commissioner who was ousted from his post said recently.[39] It has kept hundreds of eminent scientists from serving on advisory committees — with several Nobel Laureates on its now-abandoned 'blacklist.' Its concern for the profits of the pharmaceutical and pesticides industries has been notorious, so notorious that recently a bill was introduced into Congress urging the firing of all FDA officers in the highest echelon.[40] It has obligingly increased tolerances — between 1954 and 1965, fourteen of the eighteen changes were increases[41] — all the while assuring the public that no danger was involved. It extended safety claims to 2,4,5,-T – even after tests conducted for the National Cancer Institute showed that abnormal fetuses were produced by 90 and even 100 per cent of the test rats exposed to 'relatively large doses' of it, and at the lowest dose, 39 per cent.[42] It was not until the Council of European Communities slashed tolerance limits for pesticide residues for fruits and vegetables in 1968 that it finally took some action. The barring of US imports managed to effect what concern for public health could not.[43]

In some cases, reports have been altered or suppressed when they revealed adverse findings. Perhaps the most startling example was the report of the Bionetics Research Laboratory, a five-year study involving the bioassay of 130 compounds on more than 20,000 mice. A preliminary report of the findings by the research director was scheduled to be given at the annual meeting of the Society of Toxicology in March 1969 — a report that had been widely publicized as 'one of the major scientific events of the year' and one that would have far-reaching consequences since it identified some of the most commonly used pesticides as 'clearly' tumorigenic.[44]

Unable to get a copy of the report, I enlisted the aid of our Congressional representative, George E. Brown, Jr. He could not get the report either, although a few weeks later he sent me a copy of a statement by the director of the National Cancer Institute, who had commissioned the report, and a copy of a letter from US Health and Welfare Secretary Robert Finch. Both stressed the need for further studies 'in order not to mislead or alarm the American people.'[45] When the Bionetics Report was finally issued, the revised version went all out to eliminate 'anxieties' — although it did acknowledge that a number of the compounds had induced 'a significantly elevated incidence of tumors.'[46] Meantime, a substitute study commissioned by Secretary Finch had been issued — a study so bland that it advocated further investigation even of substances which had been identified as hazardous for more than a decade.[47] That is scarcely surprising considering the makeup of the commission. Co-chairmen were Dr. William J. Darby of Vanderbilt University, one of the industry's most vigorous defenders, and Dr. Emil M. Mrak, Chancellor Emeritus of the University of California at Davis. Although the 'Mrak Commission' urged the elimination of DDT and DDD in the US within two years — except when it was essential to health and welfare, Dr. Mrak later asserted that 'as far as I am concerned, DDT has not been proven harmful to the human organism.' And in the same speech to the National Agricultural Chemicals Association, he denounced the government's temporary ban on 2,4,5,-T on food crops as 'a panic-button operation.'[48]

About the spray programs conducted by various agencies and departments of the US government, when no responsible agency — including the United Nations or the American Association for the Advancement of Science — had been willing to act in relation to the defoliation of Vietnam, the SSRS did act. The report by Dr. Egbert W. Pfeiffer and Dr. Gordon H. Orians constituted a terrible indictment of Department of Defense policies and its claims that only 'safe' herbicides were being used in 'small areas.' It is chiefly thanks to the SSRS that the massive chemical war against the people and the land of Vietnam was finally halted.

That regulatory agencies at all levels have ignored their responsibilities as public servants and turned over their functions to the regulated is no secret. Partly, it is the result of the effective selling job done by industry generally and by the quality of the men at the top. The situation was described memorably by Lee Loevinger, Assistant Attorney General for Antitrust under President Kennedy:

> Unfortunately the history of every regulatory agency in the government is that it comes to represent the industry or groups it's supposed to control. All of these agencies were fine when they were first set up, but before long they became infiltrated by the regulatees and are now more or less run by and for them. It's not a question of venality either. More, the agency people consort with this or that representative of some special-interest group and finally they all come to think alike. Every company that's concerned about government control and is big enough to manage, hires a man — maybe four or five men — at anywhere from thirty to seventy thousand

dollars a year to find out what we're up to. And by God, they find out. They wine and dine agency people and get to be great friends with them. Like a lot of people without much money, some bureaucrats are impressed by being around big shots and the big life. Sooner or later, all of these agencies wind up with constituents. And they represent them damn well, too.[49]

The Loevinger theory is accurate as far as it goes. But it neglects to mention the pressures brought by legislator-merchandisers. A case in point, as regards pesticides regulation, is Jamie L. Whitten, the dapper Mississippi congressman. Chairman of the House Appropriations Subcommittee on Agriculture for the last two decades, his word on pesticides has more than the force of law around the US Department of Agriculture. And that word is summed up in his book, *That We May Live,* whose distribution was extensively subsidized by the pesticide industry. Briefly, it is that: 'The worst residue problem we have to face today is the residue of public opinion left by Rachel Carson's *Silent Spring.'*[50]

Policy-makers at USDA have shared that attitude — and well they might, if they expect to get a satisfactory appropriation from Congressman Whitten's Subcommittee. So, apparently, do the officials of the Environmental Protection Agency — the superagency set up in 1970 to see that finally the public is protected from pollutants. The glad outcries that greeted its creation quieted notably in the spring of 1971, when it was learned that Congressman Whitten had maneuvered himself into a position of controlling the purse strings for EPA, too. Subsequent events confirmed the feeling that EPA would be no different, as we saw in its back-tracking on the DDT and 2,4,5-T ban. Although the latter was outlawed in Vietnam, the EPA administrator declared that his agency did not consider 2,4,5-T or DDT and highly toxic aldrin, dieldrin, and Mirex to be 'imminent' hazards to humans and okayed their use 'pending further studies.'[51] A few months later, EPA underscored that action by raising no objections to having vital data in its planned antipollution guidelines withheld from the public. 'The public has no way of keeping us honest. That's a chance they have to take,' EPA's acting director said when questioned by reporters.

It is not a chance the public should be willing to take — particularly considering the high praises that officials of the National Agricultural Chemicals Association have had for the men who have been appointed to its leadership.[52]

Socially responsible scientists must act if the dangers of pesticide pollution as well as other forms of pollution are to be minimized. They have to some extent in Europe; but they will have to do more to educate legislators, regulators, farmers, and the lay public The pesticide industry has been profoundly disturbed by the growing caution in some countries. In 1969, for example — the last year for which complete statistics are available — pesticide exports from the US decreased 13 per cent, and the value of total pesticide exports, including inert materials in the formulations as well as active ingredients, was down 16 per cent.[53]

With hundreds of millions — even billions — of dollars at stake, the industry is obviously going to do all it can to recover its losses in sales and to expand its hitherto expanding markets.

The report for the year ending August 31, 1971, by the National Agricultural Chemicals Association offers instructive material about influencing public opinion — indeed, provides a blueprint that might be of value for counteraction.

NACA was very busy on the legislative regulatory front. It worked — successfully — to weaken or defeat nineteen bills in Congress and 251 in state and Puerto Rico legislatures to regulate, restrict, or affect marketing patterns. It worked closely with many government agencies, nationally, locally, and internationally, to establish 'a better understanding' about pesticide supervision and distribution, from the Environmental Protection Agency to *Groupment International des Associations Nationales de Fabricants de Pesticides* — 'a strong behind-the-scenes force in the development of the *reasoned approach* on use of pesticides being taken by WHO and FAO'[54] (italics added). A principal goal in dealings with regulatory agencies was to avert proposals by the Federal Trade Commission to formulate stiff rulings on the advertising of pesticides. And efforts to head off regulations for the shipment and storage of pesticides — presently there are few or none — were made by expanding and publicizing the activity of the 'Pesticide Safety Team Network.' In the eighteen months of its existence, the report noted, the Network trained forty teams and processed 139 calls for assistance in cooperation with the emergency facilities of the Atomic Energy Commission and the Office of Water Quality of the EPA.[55]

Monthly conferences were arranged with organizations like the American Farm Bureau Federation, the Chamber of Commerce of the United States, the National Farmers Association, and the National Grange to insure that members of those organizations were receiving the proper information about pesticides. NACA was a sponsor — along with the US Departments of Agriculture and the Interior, the Ford Foundation, the Rockefeller Foundation, and the National Science Foundation — of a 'significant and influential' symposium on pest-control strategies. And representatives gave sixty-seven speeches — more than in any other year since NACA's organization in 1933 — at conferences held by state and national organizations.

The 'Public Information Program,' developed by NACA in conjunction with Basford Public Relations, Inc., was little short of awesome. During the year, contacts were made with more than 1500 editors, writers, and publishers of major national magazines and radio producers and directors, and others who influence public opinion. Additionally, more than 15,000 contacts were made with media. The results and some of the details were spelled out in *Farm Chemicals:*

> Doors have been opened; some articles favorable to pesticides have resulted, some articles originally highly critical of pesticides have been modified to present a more balanced report.[56]

Consider just these few achievements highlighted in a report to the NACA Board of Directors by Basford and relayed by the magazine. The public relations agency:

(1) 'Provided material' for 'The Senseless War on Science,' an article in the March 1971 issue of *Fortune;*

(2) 'Assisted' Don Moser, *Life* magazine editor, in preparing an article in the January 22, 1971, issue of that magazine;

(3) 'Supplied pesticide material and counseled' Myles Callum, managing editor of *Better Homes and Gardens,* in preparing an editorial for the June 1971 issue;

(4) 'Provided source material' for an article in *Senior Scholastic,* a magazine widely circulated among secondary-school students;

(5) 'Provided source material' resulting in 'a highly favorable column' on pesticide controversy to Walter Trohan, Chicago *Tribune* syndicated columnist;

(6) 'Provided material' to the Newspaper Enterprise Association that resulted in 'a highly favorable syndicated editorial' on pesticides sent to 150 newspapers throughout the country, and later 'counseled' NEA science editor Dave Hendrin on a pesticide article released to 700 newspapers — an article that had 'a softer, better rounded approach.'

Radio and television success was even more impressive. A 'public service' television one-minute spot distributed in October 1970 to 240 stations throughout the country resulted in more than thirty-five hours of broadcast time, 2000 separate telecasts in sixty-seven cities. It is estimated that the showing by New York City stations alone was worth 30,000 dollars in broadcast time. A second one-minute spot distributed in February 1971 by Basford — 'Preserve Our Natural Resources' — was held to be even more successful. Between the release date and

111

August, when the NACA Annual Report was prepared, it had logged forty-eight hours of broadcast time, representing about 2800 individual television showings. One of the sixty-second spot announcements for radio, 'Panorama,' received 197 plays on sixty-five stations in thirty-three states within a few months. A thirty-second TV 'news spot' featuring Dr. Borlaug was telecast in twenty-seven market areas, including Los Angeles, San Francisco, and Cleveland; that 'news spot' was shipped to ABC-TV affiliates in forty foreign countries. The major message by Borlaug was the assertion that 'either we use agricultural chemicals and use them wisely in the right amounts to produce the food we need or we will all starve.'

One of the 'major elements' of the 1972 public-relations program, it was noted, will be the distribution of a fifteen-minute film documentary: 'Norman Borlaug — Revolutionary.' At least 120 prints of the film had been placed in distribution channels by August 1971; it was entered for showing at two film festivals. Release of the Borlaug film will be directed to community, civic, and service groups; but 'prime target audience' will be the secondary school system.

With a Nobel Prize winner as hero, the film should do much to open wider the doors of the US school system. The first major assault on schools in the spring of 1971 effected an impressive entry. For example, an initial mailing to 16,000 teachers of NACA 'informational' materials resulted in a response from more than 1200 of them. As a consequence, more than 450 showings of propesticide films were arranged; 35,000 books reprints, and posters were distributed along with 135,000 cartoon booklets. The cartoon booklet, *The I Learn about Bugs and Other Things Coloring Book,* is as long on propaganda as it is short on biology, chemistry, and reason and human feeling. The only creatures portrayed — with the exception of a house cricket — are those calculated to stir a loathing of living things: roaches, black widow spiders, fleas, rats, and mice. The implication is that every small creature should be wiped out — preferably by that wildly smiling and waving pest-control operator, who is to be colored 'very important.'[57] The restaurant, the house, and other items pictured are to be colored 'clean and healthy' after structures and contents have been dosed with pesticides.

Opponents of the indiscriminate use of pesticides have much less opportunity to get the message to listeners, viewers, and readers — as I have every reason to know. For example, although *Unfit for Human Consumption* had an excellent response from scientists, including Dr. Ralph Lapp and Dr. Alice Mary Hilton; from consumer advocates like Jerry Voorhis, former president of the Cooperative League of the USA and the International Cooperative League, James Turner of Ralph Nader's center and author of *The Chemical Feast;* from 'environmental' legislators like Senator Alan Cranston and Senator Gaylord Nelson, it was not given much notice in the popular press. (That is not exceptional. Ralph Nader's *Unsafe at Any Speed* was not mentioned by any major newspaper or magazine until he made front-page headlines because of his harassment by General Motors after testifying before the Ribicoff Senate Subcommittee.) A friend of twenty-five years spelled out the reasons for me: As city editor of one of the largest metropolitan daily newspapers in the United States, he sent me a thank-you note for the publisher's copy of the book, along with a press release he had just received containing the utterances of Dr. Max Tishler, president-elect of the American Chemical Society, who was demanding among other things: 'Is it reasonable to insist on a cleaner environment if in attaining it communities are demoralized and their citizens become welfare recipients?'[58]

My friend wrote of the press release that he hoped it would 'delight' me as it did him, and then said:

> I wish there were some way I could sneak a piece of promotion past my editors here, but they are guys who keep toying with stories like 'Be Glad You're Loaded with Strontium-90,' and the like, so I do not believe there is much hope. Keep up the good fight.

He had told me earlier about his experiences as editor of another paper. Even before the publication of *Silent Spring,* he had been very concerned about the effects of pesticides on health and the environment and had assigned a reporter to look into it. The reporter's six-month investigative article was permanently filed. 'This stuff is dynamite' was the reason given by the editor's superiors.

But there are ways in which socially responsible scientists can get the message across. They can demand of radio and television stations equal time to answer the comments of such pesticide advocates as Dr. Robert White-Stevens, one of whose programs on pesticides was aired in two New York stations in a single week and on nineteen stations around the country the following week.[59] They can write 'Letters to the Editor' after the manner of Dr. J. Gordon Edwards, Dr. Mitchell Zavon, and other pesticide advocates whose stout defenses of DDT and other pesticides are wont to appear in newspapers and magazines that have not exhibited the proper respect for the arsenal. Most importantly, socially responsible scientists have an obligation as well as an opportunity to insure that the schoolrooms are not permitted to become indoctrination centers for pesticides.

The problem is serious, and the hour is late.

NOTES AND CITED REFERENCES

1. Dr. van den Bosch's statement was made on January 30, 1969, when he was testifying at court hearings in Bakersfield, California, on the petition of the United Farm Workers Organizing Committee to force Kern County Agricultural Commissioner Seldon G. Morley to release pesticide application records. Although the records are 'public records,' the Commissioner refused to release them on the grounds that they contained 'trade secrets' — a connection that was upheld by the judge.

2. 'Who Dug for Dirt on Earth Day?' *Time* Magazine, April 26, 1971, p. 23. Senator Muskie made public an eleven-page intelligence report, written on FBI stationery, covering the 1970 Washington Earth Day rally. In the report, due note was taken of the fact that the Senator himself had given 'a short antipollution speech.'

3. William Tuohy, 'Scientist Raps "Hysterical Environmentalists," ' Los Angeles *Times,* November 9, 1971. In the speech in Rome, Dr. Borlaug attacked even more vigorously than in Washington those who opposed the unrestricted use of pesticides. He said that the 'current vicious, hysterical propaganda campaign against the use of agricultural chemicals' had its genesis in Rachel Carson's *Silent Spring,* which he labeled 'a diabolic, vitriolic attack on the use of pesticides,' Mr. Tuohy reported.

4. 'A Charge of "Hysteria" Over Farm Chemicals,' Los Angeles *Herald Examiner,* October 8, 1971.

5. Philip Fradkin, 'Biologists Blame Ecologists, Not DDT, for Pelican Decline,' Los Angeles *Times,* January 20, 1972. The state college scientists making the charge were Dr. J. Gordon Edwards, professor of entomology at San Jose State College, and Richard E. Main, curator of the college entomological museum.

6. *Agricultural Newsletter,* Extension Division, Public Relations Department, E. I. du Pont de Nemours & Company, Inc., September-October 1948, p. 83. That article makes an interesting contrast with those on DDT published in the January-February 1946 *Newsletter,* pp. 2—13.

7. Dr. Francisco Bravo, quoted *El Malcriado,* February 15, 1969, p. 15. Dr. Bravo, a member of the State Board of Agriculture, accused the chemical industry of 'considerable price gouging' and 'possible collusion' to fix prices.

8. For example, one of the narrow-range pesticides that has been a bestseller for years in those countries is Isolane, which was produced in the 1950's by Geigy, the company that patented DDT and was for long its major manufacturer. According to a Geigy apologist, the reason that more primitive pesticides are used in the USA is that 'in most cases the market for a specific is too small to warrant large-scale production.' Bill Ballantine, *Nobody Loves a Cookroach,* Boston, Little, Brown, 1967, p. 313.

9. Ray F. Smith and Robert van den Bosch, 'Integrated Control,' *Pest Control: Biological, Physical, and Selected Chemical Methods,* ed. Wendell W. Kilgore Richard L. Doutt, New York, Academic Press, 1967, pp. 328—334.

10. 'The Melancholy Addiction of Ol' King Cotton,' *Natural History,* December 1971; pp. 86—90. Dr. van den Bosch also describes the disastrous economic effects of broad-spectrum pesticides on the California cotton crop in *Chemical Fallout: Current Research on Persistent Pesticides,* ed. Morton W. Miller and George C. Berg, Springfield, II., Charles C. Thomas, 1969, p. 110.

11. Sanitary measures have been so effective in curbing cotton pests in some areas in the USA that Texas and New Mexico have laws requiring farmers to shred stalks and plow them under — a measure that enhances soil fertility.

114

Yet so bemused are farmers by pesticide sales promotion that 'all along the path of the pink bollworm, there has been a tendency on the part of some cotton farmers to object to certain cultural control regulations.' 'Fighting the Pink Bollworm,' *Cotton Farming,* January 1968 p. 16.

12. George Getze, 'Mosquitos Build Resistance to Pesticides Replacing DDT,' *Los Angeles Times,* June 21, 1971.

13. Early warning of this by Dr. E. Gorton Linsley, dean of the College of Agricultural Sciences at the University of California, Berkeley *(Science News,* June 21, 1969, p. 599) were confirmed in the summer of 1971: 'Mosquitoes: Powder Keg for California,' Los Angeles *Herald-Examiner,* July 28, 1971.

14. A. W. A. Brown, *Insecticide Resistance in Arthropods,* Geneva, World Health Organization, 1958, p. 104.

15. Stewart Udall, former US Secretary of interior, 'Public Policy and Pest Control,' *Scientific Aspects of Pest Control,* Washington, D. C., National Academy of Science-National Research Council, 1966, p. 430.

16. 'Technology-Ecology Conflict Perils Mankind, Scientists Says,' *Los Angeles Times,* November 27, 1969. (The Borneo episode had been reported earlier in *Science Digest,* September 1968.)

17. R. A. M. Case, 'Toxic Effects of DDT in Man,' *British Medical Journal,* December 15, 1945, pp. 842–845. V. D. Wigglesworth, 'A Case of DDT Poisoning in Man,' *British Medical Journal,* April 14, 1945, p. 517.

18. *Chemicals in Foods and Cosmetics, Hearings before the House Select Committee to Investigate the Use of Chemicals in Foods and Cosmetics, House of Representatives, Eighty-Second Congress.* Washington, D. C., U. S. Government Printing Office, 1952–1953, Part 2, pp. 948—963. Dr. Bernard Krohn submitted as part of his testimony an article written by him and Dr. Francis M Pottenger, Jr. The two doctors told that they had recorded in their clinic in California during a single year more than 100 cases of a syndrome resulting from the widespread use of chlorinated hydrocarbons — a syndome of hepatic and neurologic damage and sometimes death.

19. Dr. William Coda Martin, who repeated the substance of his testimony in an introduction to Leonard Wickenden's *Our Daily Poison,* New York, The Devin-Adair Company, 1956, pp. 2—21.

20. An extended discussion of the Hueper conflict with Dr. Wayland J. Hayes, Jr., is contained in Laura Tallian's *The Pesticide Jungle* (privately printed, 1966, Box 34, Phillipsville, Calif. 95559), p. 25.

21. See transcript of hearings conducted by the Wisconsin Department of Natural Resources, 1969, on the petition of the Environmental Defense Fund to ban DDT in that state. Dr. Risebrough's arguments also contained in *Chemical Fallout: Current Research on Persistent Pesticides,* p. 21.

22. Louis McLean quoted, *The Connecticut Conservation Reporter,* May-June 1969, p. 3.

23. 'The Effect of Known Repeated Oral Doses of Chorophenothane (DDT) in Man,' *Journal of the American Medical Association,* October 27, 1956, p. 897. (Actually, the paper had three authors: Dr. William F. Durham and Cipriano Cueto, Jr., also signed it — although Hayes is given sole credit in most discussions.)

24. *Newsweek,* August 25, 1969, p. 71.

25. Wayland J. Hayes, Jr., *op. cit.,* p. 893.

26. *Chemicals in Food Products, Hearings before the House Select Committee to Investigate the Use of Chemicals in Foods and Cosmetics, Eighty-First Congress.* Washington, D. C., US Government Printing Office, 1951, p. 389.

27. *Interagency Coordination in Environmental Hazards, Hearings before the Subcommittee on Reorganization of the Committee on Government Operations, U. S. Senate, Eighty-Eighth Congress.* Washington, D. C., US Government Printing Office, 1963, Part 2, pp. 484—497.

28. Human Relations Agency, California Department of Public Health, *A Report to the 1970 Legislature on the Effects of the Use of DDT and Similar Pesticides on Human Health and the Environment,* December 10, 1969, Table 1, Section VI (xerox copy).

29. *Ibid.*

30. M. C. Warren, *et al.,* 'Clothing-Borne Epidemic,' *Journal of the American Medical Association,* CLXXIV: 266, 1963. Also, 'The Dead Mosquitoes,' *Annals of Medicine* series, *The New Yorker,* October 11, 1969, pp. 123—142.

31. Statement of Jerome B. Gordon before US Senate Subcommittee on Migratory Labor, August 1, 1969. A typescript of Mr. Gordon's statement was given to me by Senator Walter F. Mondale. See also, *Migrant and Seasonal Farmworker Powerlessness, Hearings Before the Subcommittee on Migratory Labor of the Committee on Labor and Public Welfare, US Senate, Ninety-First Congress,* Washington, D. C., Government Printing Office, 1971.

32. *El Malcriado,* November 15—30, 1969, pp. 3 and 7.

33. S. Gershon and F. H. Shaw, *The Lancet,* June 1961, pp. 1371—74.

34. Dr. Neville Grant, 'Legacy of the Mad Hatter,' *Environment,* May 19, 1969, pp. 18—24. Also in that issue, Goran Lefroth and Margaret E. Duffy, 'Birds Give Warning,' p. 12.

35. *Safe Use of Pesticides in Public Health; Sixteenth Report of the WHO Expert Committee on Insecticides,* World Health Organization Technical Report Series, No. 356, Geneva, 1967, p. 5.

36. 'UNICEF Shipments of DDT Come Under Cloud,' *Los Angeles Times,* November 20, 1969.

37. *Deficiencies in Administration of Federal Insecticide, Fungicide, and Rodenticide Act: Hearings Before a Subcommittee of the Committee on Government Operations, Ninety-first Congress, First Session,* Washington, D. C., US Government Printing Office, 1969, pp. 151—159.

38. *Deficiencies in Administration of Federal Insecticide, Fungicide, and Rodenticide Act, Eleventh Report by the Committee on Government Operations, Ninety-first Congress, First Session,* Washington, D. C., US Government Printing Office, 1969, p. 71. The quotation is from an article by Kenneth D. Quarterman, then head of the Communicable Disease Center in Savannah. It further states: 'It was decided also that the Government should not seek to obtain a domestic patent on the invention because prior publications on the invention by the Government scientist who developed it are deemed sufficient protection against the prosecution of a successful patent application by a later inventor.'

39. Dr. Herbert L. Ley, Jr., quoted in the *Los Angeles Times,* January 1, 1970. In an interview of industry pressures, he said: 'Some days I spent as many as six hours fending off representatives of the drug industry.'

40. Senate Bill 983 ('the Consumer Product Safety Bill'), introduced in 1971 by Senator Frank E. Moss, proposed that the FDA be replaced with a more effective Consumer Safety Agency. FDA employees in grades up to GS-11 would be transferred to the new agency, while those in GS-12 or higher would be swept out for having consistently 'made a mockery of protecting the public.'

41. Booth Mooney, *The Hidden Assassins,* Chicago, Follett Publishing Company,

1966, p. 111. Dr. Paul Ehrlich noted in *The Population Bomb*, New York, Ballantine Books, 1968: 'The setting of tolerances by the FDA is much too open to error (as can be seen by repeated readjustments) and the power available to enforce tolerances is completely inadequate,' (pp. 122—123).

42. Thomas Whiteside, *The Withering Rain, America's Herbicidal Folly,* New York, E. P. Dutton, 1971, p. 43.

43. 'Unofficial Translation of Proposed Pesticide Regulation of the Common Market,' *Report to the California-Arizona Citrus Industry* by D. R. Thompson, European representative, December 24, 1968. In that year, the FDA lowered DDT residue tolerance on 36 fruits and vegetables from 7 ppm to 3.6 ppm. The tolerance in Europe is only 1 ppm. There were other significant differences: i.e. the European tolerance for Captan, 15 ppm; US 100 ppm, etc.

44. The full story of the Bionetics report is contained in *Effects of 2,4,5-T on Man and the Environment, Hearings before the Subcommittee on Energy, Natural Resources, and the Environment of the Committee on Commerce, U. S. Senate, Ninety-first Congress,* Washington, D. C., US Government Printing Office, 1970.

45. Copy of a letter dated May 6, 1969.

46. *Interim Report on Studies of Pesticides and Other Agricultural Chemicals,* National Cancer Institute, 1969, pp. 2—3.

47. *Recommendations and Summaries of the Secretary's Commission on Pesticides and their Relationship to Environmental Health,* Washington, D. C., Department of Health, Education, and Welfare, November, 1969, p. 6.

48. 'Industry Must Speak Up, Mrak Tells NACA,' *Farm Chemicals,* November 1971, p. 22.

49. James Bishop, Jr., and Henry W. Hubbard, *Let the Seller Beware!* Washington, D. C., The National Press, 1969, p. 88.

50. Representative Whitten quoted, *Conservation Foundation Letter,* May 5, 1969, p. 6.

51. *Time,* April 12, 1971.

52. According to a report in *Farm Chemicals,* April 1971, Parke Brinkley, NACA President, said: 'To date, I'm more impressed with his performance than any other Nixon appointee.' Brinkley was referring to William Ruckelshaus, head of the Environmental Protection Agency. According to the editor of the magazine: 'Ruckelshaus is fast earning the reputation among industry leaders for objectiveness and fairness. Ruckelshaus gained more support when he refused to buckle under the pressure of environmentalists who were demanding an immediate ban on DDT and 2,4,5-T as imminent hazards to human health'

53. Agricultural Stabilization and Conservation Service, United States Department of Agriculture, *Pesticide Review 1970,* p. 1.

54. 'GIFAP Works Behind the Scenes,' *Farm Chemicals,* September 1971, p. 26.

55. *National Agricultural Chemicals Association Annual Report,* Year Ending August 31, 1971, p. 24.

56. 'NACA's Constructive Year,' *Farm Chemicals,* September 1971, pp. 22—24. (The NACA Annual Report did not include the names of writers and publications 'counseled' by the Basford agency.)

57. The cartoon booklet I received was distributed by the industry at the University of California at Santa Cruz Earth Day observance in 1971.

58. Press Release issued to science editors by Sherwood Ross Associates, May 1971. The release called attention to Dr. Tishler's article in the June 5, 1971, issue of *The Saturday Review.*

59. *Farm Chemicals,* September 1971, p. 22.

Against Pollution: Radiation

E. J. STERNGLASS

Nuclear Radiation
and Human Health

The present paper will address itself to the evidence that points to the conclusion that low-level radiation from nuclear fission products in the environment, such as those released by nuclear explosions and power reactors, may already have produced serious effects on the health of the world's population. These effects are far beyond those ever believed possible when our present radiation standards were originally formulated and adopted.

Before discussing the latest evidence in detail, I should like to review very briefly the nature of the early discovery that low-level radiation may have detectable effects, and also to summarize the difficulties that have existed until now in unequivocally relating the observed effects to the action of nuclear fallout.

The earliest indication that low-level radiation could produce serious effects in man came from the studies of Dr. Alice Stewart at Oxford University in 1958 showing that mothers who had received a series of three to five pelvic x-rays during pregnancy had children who were almost twice as likely to develop leukemia and other cancers before age ten than mothers who had no pelvic x-ray examinations.[1]

This work was independently confirmed in 1962 in a major study involving close to 800,000 children born in New York and New England by Dr. Brian MacMahon of the Harvard School of Public Health.[2] Using these two sets of data, it was possible to show that there appears to exist a direct, straight-line relationship between the number of x-rays given to a pregnant woman and the probability that the child will subsequently develop leukemia, and that there is therefore no evidence for the existence of a safe 'threshold level' below which no additional cancers are produced, down to the relatively small dose from a single x-ray. Furthermore, the magnitude of the x-ray dose to the developing fetus in utero from one such x-ray was comparable with the dose normally received in the course of two to three years of natural background radiation, or from the fallout produced in the course of the 1961–1963 test series, namely 0.2–0.3 rad.[3]

121

These early findings have since been confirmed by the results of Dr. Stewart, published in June of 1970.[4] This extensive study, based on over 7000 children born in England and Wales between 1943 and 1965 who developed leukemia or other cancers, gave the result that for 1 rad to a population of a million children exposed shortly before birth, there were an extra 300—800 cancer deaths before age ten with a mean number of 572 ± 133 per rad. For a normal rate of incidence of about 700 cases per million children born, this means that only 1.2 rads (1200 mr) are required to double the spontaneous incidence. Furthermore, Dr. Stewart's study showed that when the radiation exposure took place in the first trimester, the excess risk of cancer increased fifteen times. This means that a dose of only some 180 mr was found to double the normal cancer risk for the early embryo, much less than the presently permitted 500 mr annual dose to any member of the general population.

It was therefore possible that studies of large populations of children exposed to known incidents of localized fallout in a given area might show detectable increases in leukemia some years later. Such a localized 'rain-out' was pointed out by Ralph Lapp[5] as having taken place in Albany-Troy, New York, in April of 1953 following the detonation of a 40 kiloton bomb in Nevada. An examination of the data on leukemia incidence published by the New York State Department of Health showed that when plotted by year of death there was a clear increase in the number of cases per year among children under ten years of age at death from about 2—3 to as many as 8—9 per year some 6—8 years after the arrival of the fallout, exactly the same delay in peak incidence as observed in Hiroshima and Nagasaki. Furthermore, the peak contained many children who were not even conceived until a year or more after the arrival of the fallout, suggesting for the first time the existence of an effect prior to conception.

Because of the relatively small number of cases in Albany-Troy, it was difficult to draw absolutely firm conclusions, and so the situation for New York State as a whole was examined. Again, peaks of leukemia incidence were clearly present some 4—6 years after known atmospheric tests in Nevada; this greatly strengthens the initial observations for Albany-Troy alone.[6]

Following the arrival of the fallout in Albany-Troy in 1953, there was also a drastic slow-down in the steady decline of fetal mortality or stillbirths in that area. Following up this unexpected finding, the fetal and infant mortality statistics for New York State as a whole were examined, followed by those for California and other states. The same slow-down in the decline or even renewed rises in the mortality rates existed to varying degrees depending on the amount of fallout in the milk,

122

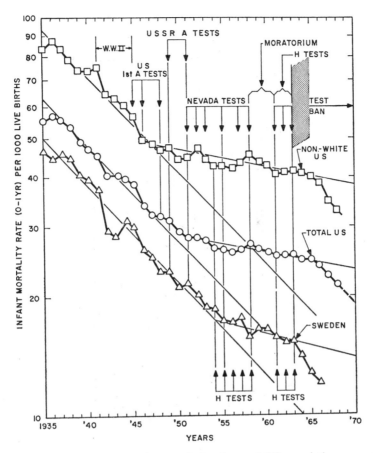

Figure 1. Infant mortality rate (0–1 yrs) for the total US population, non-white population, and Sweden. The decline that began in 1965–1967 has continued in the USA so that by 1970, the rate for the total population reached 20 per 1000 births (US Vital Statistics).

beginning in the early fifties, the declines resuming only 2–4 years after the end of atmospheric testing in 1962.[6,7] For the USA as a whole, the data are shown in Figure 1, where the infant mortality rates for both the total population and the non-white population have been plotted together with the data for Sweden. It was then drawn to our attention that Dr. I. M. Moriyama of the US National Center for Health Statistics had drawn attention to the leveling trend in the USA beginning in about 1951 as early as 1960[8] and that he had, in fact, suggested the possibility that similar upward changes of mortality for all age groups

123

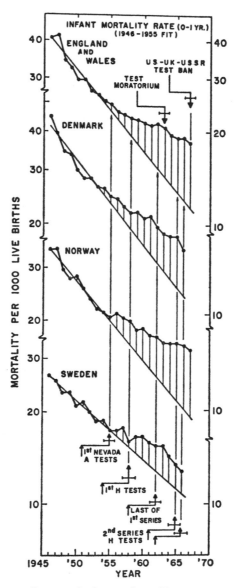

Figure 2. Infant mortality trends for northern European countries after World War II. Note onset of upward deviations peaking some 3–5 yrs after major test series. Least square fits to 1946–1955 trend.

might be connected with the sharp rises in environmental radioactivity from nuclear testing.[9]

Since then, we have extended our studies to other countries in the

124

world, and especially in northern Europe, which received the fallout from the Nevada tests in its northeasterly drift across the Atlantic, and the same patterns of slowdown followed by a renewed decline of infant mortality were found, as shown in Figure 2. At the same time the leveling trends were much less pronounced in countries like Canada, France, and Italy that were to the north or south of the path of the Nevada fallout on its northeasterly course across the USA, and the North Atlantic so that they did not receive the short-lived activity.

 We have since established high degrees of correlation between the increases in infant mortality above the declining base-lines, and the measured Strontium-90 levels in the milk, and in the bone of children and young adults for all the nine states of the Public Health Service's Raw Milk Network, for which data are available back to 1957–1958.[10] These correlations suggest that as many as 400,000 infants 0–1 year old in the USA alone may have died as the result of nuclear testing.

These results are so startling and so unexpected, that they have encountered considerable skepticism, primarily because the technique of trend-analysis as used first by Moriyama to calculate 'excess deaths' above normal expectations for all age groups in the USA was based on the expectation of a steadily declining infant mortality at least until levels are reached equal to those that had already been attained in other medically advanced nations of the world such as Sweden (see Figure 1). Such an assumption, however, is justified by the fact that in New Mexico, after the initial test in 1945, there was a return to the same steady decline, determined by the computer fit to the 1935–1950 period, due to the low rainfall and therefore low levels of fallout in the milk after 1950 — when nuclear testing was moved north to Nevada. Furthermore, the most recent data on infant mortality show that in a number of rural states such as Vermont, Maine, and Nebraska infant mortality rates have declined very sharply, reaching the levels predicted on the basis of the 1935–1950 rate, as illustrated in Figure 3 for Maine. Nevertheless, there is some degree of arbitrariness inherent in using any given period of declining mortality rates as a baseline, and it is therefore important to find other data not subject to the same criticism.

Fallout and congenital malformations

Such data exist in the case of childhood deaths associated with congenital malformations such as Down's Syndrome, microcephaly, and congenital heart defects. For this particular category of infant and childhood deaths, there has been only a slight downward trend over the last twenty years, since neither the introduction of new antibiotics nor medical care methods, nor the gradual improvement in diet has had

125

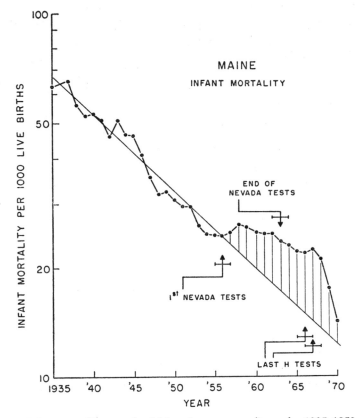

Figure 3. Infant mortality rate for Maine. Least square fit to the 1935–1950 trend (US Vital Statistics).

significant effects on these mortality rates. As a result, there is here no need to extrapolate a rapid downward trend, and one has, for every state and many foreign countries, a well-established, nearly horizontal baseline to the onset of nuclear testing in 1945. Furthermore, congenital malformations are well known to be capable of being caused by relatively low levels of radiation in animals, and recent studies on such conditions as mental retardation published by the Scientific Commission of Radiation[11] have established that small amounts of radiation during certain critical periods of embryonic development and organ formation can produce detectable effects in children.

We therefore examined the incidence of deaths among congenitally defective children in relation to children who died of accidents as a control group at various distances from the Nevada test site, where relatively high local fallout was known to have occurred in a number of

126

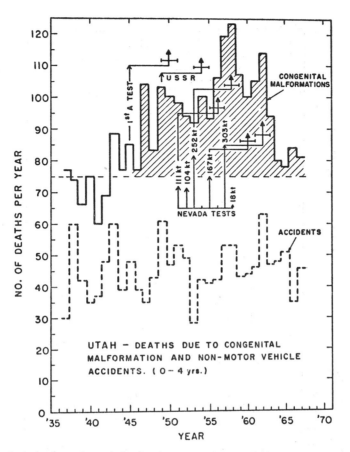

Figure 4. Annual number of deaths due to congenital malformations in Utah for children aged 0–4 yrs, compared with annual deaths due to non-motor vehicle accidents as a control (US Vital Statistics).

instances, documented both by the AEC[12] and independent studies by scientists at the University of Utah[13] and the St. Louis Center for Nuclear Information.[14]

As an example, in Figure 4 is shown the annual number of deaths of congenitally defective children 0–4 years old in Utah directly east of Nevada and therefore generally downwind from the test site as taken from the published figures in the US Vital Statistics, together with the deaths in this age group due to accidents other than those involving automobiles. It is seen that the average number of deaths of congenitally defective children per year in the pretesting period 1937–1945 stayed relatively constant at about 75 cases per year, but that it rose to a

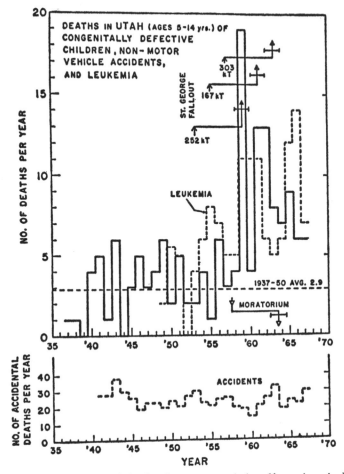

Figure 5. Annual number of deaths due to congenital malformations in Utah for children aged 5–14 yrs, compared with deaths due to accidents (US Vital Statistics).

peak of 123 cases per year in 1958, some five years after a particularly heavy fallout incident in 1953, returning close to the pretesting rate of 80 per year five years after the end of atmospheric tests in Nevada. Such a rise and decline while accidental deaths remained constant is clearly not explainable in terms of gradual rise in the number of births per year. Altogether, there seem to be some 480 children likely to have died of congenital malformations in Utah above expectations, based on a comparison with the number of accidental deaths since the onset of nuclear testing in 1945.

An even more striking peak in deaths of congenitally defective chil-

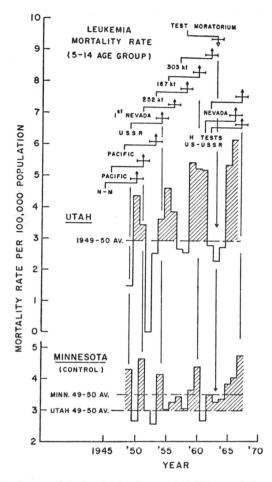

Figure 6. Annual rate of leukemia deaths per 100,000 population aged 5–14 yrs for Utah near the Nevada test site compared with Minnesota as control. Major test series are indicated, together with the 4–6-yr delay expected for leukemia (US Vital Statistics).

dren relative to the number of accidental deaths took place in the 5–14-year age group shown in Figure 5 for the case of Utah, which includes children who received radiation from the milk and food some time after birth. Again, a 4–6-year delay is seen to occur between exposure and death, quite similar to the case of Hiroshima and Albany-Troy, corresponding to the fact that children born congenitally defective are much more prone to develop leukemia with its 4–6-year delay of peak incidence.

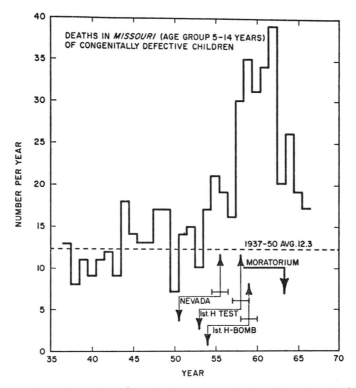

Figure 7. Annual number of deaths due to congenital malformations in Missouri, age group 5–14 yrs.

The rate of leukemia deaths for all children in the age group 5–15 years, which was shown by Stewart[1] and MacMahon[2] to reflect the effects of perinatal irradiation most strongly, is plotted in Figure 6 for Utah. It is seen that statistically significant peaks occurred some 4–6 years after known tests had deposited fallout in Utah, apparently affecting the infants both prior to and after birth. Furthermore, the relative increases were higher than those observed in Minnesota (control), as is to be expected from the great proximity to the test site. Thus, the effects are observed for both annual numbers and rates per 100,000 population.

Other examples of similar though lower relative rises in deaths among children born defective are seen in the plots for states generally to the east of New Mexico and Nevada, such as Missouri, Georgia, and Texas (see Figures 7–9). In the case of Texas, leukemia deaths for all children aged 5–14 years have also been included, again showing a parallel rise

130

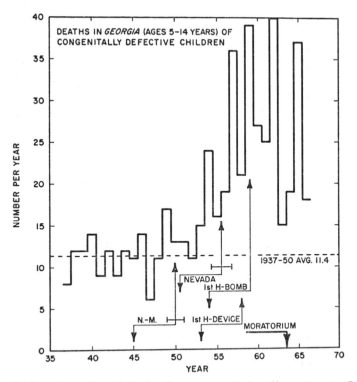

Figure 8. Annual number of deaths due to congenital malformations in Georgia, age group 5–14 yrs.

with deaths of congenitally defective children. No other explanation of these striking rises and declines in mortality rates is known.

The reason why such unexpectedly large effects of fallout should be observed — when radiation levels were believed to be so low as to be regarded as completely safe — is evidently connected with the much greater sensitivity of the embryo and infant compared with the adult.

Furthermore, the severity of the effects is also connected with the biological concentration of certain isotopes in the food chain, mainly via the milk, which was not widely recognized at the time when the tests were begun. Another reason is the selective concentration, in various critical organs of the human body, of certain isotopes whose biological consequences were not fully appreciated for the sensitive developmental phase of the early embryo and fetus.

Thus, experimental studies on laboratory animals by Dr. Walter Müller published in 1967[15] suggest that strontium-90 and other alkaline-earth elements long known to seek out bone may also produce biological

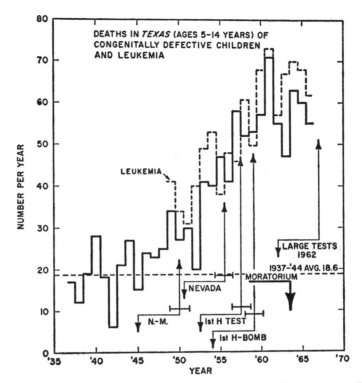

Figure 9. Annual number of deaths due to congenital malformations in Texas, age group 5–14 yrs. Annual number of leukemia cases also shown for same age group.

and possibly genetic effects through their daughter elements, such as yttrium-90, into which they decay and which are known to preferentially concentrate in such vital glands as the pituitary, the liver, the pancreas, and the male and female reproductive glands.[16,17]

In any case, we are apparently confronted with still another unanticipated biological concentration effect similar to the surprises we received when we discovered the special hazard of iodine-131 going to the infant thyroid and strontium-90 and -89 going to the bone via the originally unsuspected pathway of milk produced by cows grazing on contaminated pastures.

Infant mortality and releases from nuclear reactors

Similarly unanticipated effects on the developing embryo and infant may have taken place as a result of fission products released from nu-

132

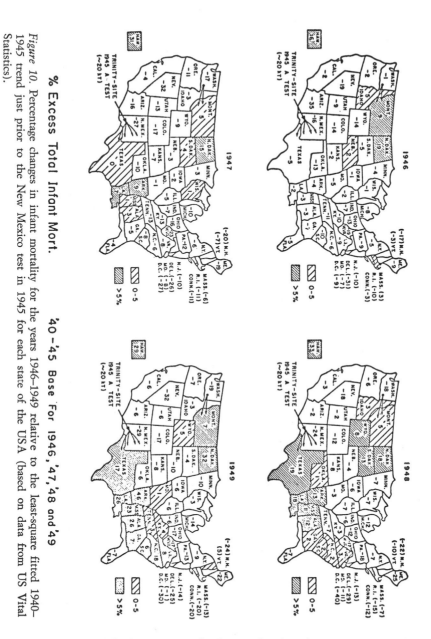

Figure 10. Percentage changes in infant mortality for the years 1946-1949 relative to the least-square fitted 1940-1945 trend just prior to the New Mexico test in 1945 for each state of the USA (based on data from US Vital Statistics).

clear reactors and fuel-processing facilities; this first became apparent in the course of our state-by-state study of infant mortality changes following the first nuclear weapons test in New Mexico in 1945.

As shown in Figure 10, each map for the four years following this

133

test showing the percentage changes relative to the trend for the previous five years indicated an upward change not only in infant mortality directly to the east and northeast of New Mexico, but also in the states to the east of the Hanford plutonium production facility in the State of Washington.

Not only were the Hanford Reactor and plutonium production facilities operating at very high levels from 1944, releasing into the environment the rare gases that could not be readily trapped, but on a number of occasions there were serious accidents, in the course of extracting the plutonium from the irradiated uranium fuel elements by chemical techniques, when fuel elements burst into flames and discharged large quantities of fission products into the environment.[18]

The infant mortality changes were greater in North Dakota than in dry Idaho and Montana, just as they were greater in Arkansas and Louisiana compared with dry Texas closer to the test site in New Mexico. This fits the well-known fact that 90 per cent of the fine tropospheric fallout comes down with the rain, since the line of heavy rainfalls passes down through the center of the United States just to the West of the Mississippi from North Dakota in the north to eastern Texas in the south.

This interpretation is further confirmed by a more detailed analysis of infant mortality changes in the counties near the Hanford plant before and after it went into operation between 1943 and 1945. As can be seen from the bar graph in Figure 11, the country containing the plant as well as those immediately adjacent to the east and south showed sharp rises in infant mortality up to 150 per cent, while the more distant control counties, namely those in which water-sampling stations were subsequently established, either rose less than 10 per cent or actually declined between 1943 and 1945.

A similar pattern of increased infant mortality has now been observed around three commercial nuclear power reactors of the boiling-water type (BWR), in which the single-coolant loop design does not permit as tight a containment of fission products leaking out of corroded fuel elements as in the naval-submarine type pressurized water reactor.

As described in recent publications of the Bureau of Radiological Health,[19] these reactors have emitted as much as 800,000 curies of fission and neutron-activation products in the form of gases per year,[20] compared with as little as 0.001 curie per year for the prototype Pressurized Water Reactor at Shippingport, Pennsylvania.

The first of the BWR's studied is the Dresden Reactor located near Morris, Illinois, in Grundy County, some 50 miles southwest of Chicago. Since close to two-thirds of the population of Illinois live within

134

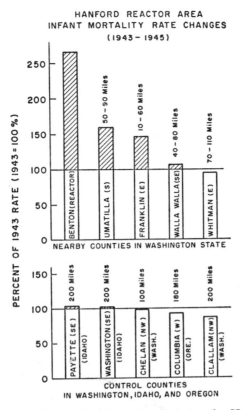

HANFORD REACTOR AREA
INFANT MORTALITY RATE CHANGES
(1943 - 1945)

Figure 11. Percentage changes in infant mortality near the Hanford Reactor in Washington before and after onset of operations in 1944. Control counties are those where water-sampling stations were placed (based on data in US Vital Statistics).

a radius of some sixty miles from this reactor, one might expect to find detectable changes in infant mortality for Illinois as a whole, relative to other nearby states, that correlate with the rises and declines of emission when fuel elements are changed.

That this appears in fact to have taken place is illustrated by the plot of infant mortality for Illinois compared with Ohio some 200 miles to the east for the period 1959–1968 in Figure 12. It is seen that, while during the time of Nevada testing, Ohio and Illinois showed the same infant mortality, within a few years after the end of testing Ohio began a steady decline, whereas Illinois showed a peak highly correlated with the peak of gaseous emissions between 1964 and 1967.

The degree of correlation may be judged from Figure 13, where the difference in infant mortality rates between Illinois and Ohio has been

135

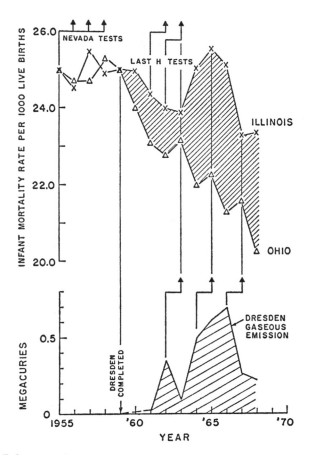

Figure 12. Infant mortality in Illinois compared with Ohio for the period 1955–1968. Also shown are the annual gaseous releases from the Dresden Reactor (US Vital Statistics).

plotted against the annual gaseous discharges. The correlation coefficient is 0.865, and the t-test of significance gives t = 4.565, which for the seven degrees of freedom gives P $\ll$ 0.01.

As in the case of Hanford, it is of interest to see whether the effect can also be detected in the nearby states to the east, the direction in which the prevailing winds and weather patterns move. As seen in Figure 14, the infant mortality rate for nearby Indiana does indeed fall exactly between that for Illinois and Ohio on the other side of Indiana after the testing in Nevada ended and the discharges from the Dresden Reactor produced significant external doses, comparable with those from distant tests.

136

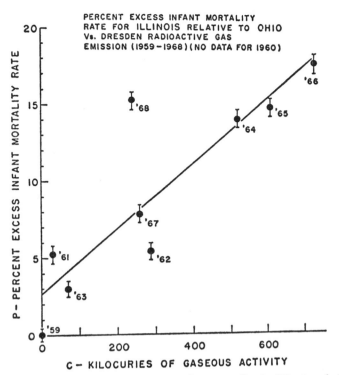

Figure 13. Correlation plot for the excess infant mortality in Illinois relative to Ohio vs. the annual average quantities of gaseous releases from the Dresden reactor. Least square fitted line shown.

Likewise in Michigan, just to the north of Indiana, infant mortality began to fall consistently between Illinois and more distant Ohio when the general decline began after the end of nuclear testing in 1963 (see Figure 15).

One would also expect on the basis of this hypothesis that a state far to the northwest of Illinois and therefore upwind would show an even more rapid decline after fallout from weapons testing decreased. That this is in fact the case is seen for the case of North Dakota compared with Illinois in Figure 16.

The rates for Illinois and North Dakota seem to have been identical during the period of heavy Nevada testing and plutonium production at Hanford prior to 1964, despite the great difference in ordinary air pollution and socioeconomic character of the two states. But after the end of nuclear testing by the USA and USSR, North Dakota declined rapidly from nearly 25 per 1000 births to under 17 per 1000 by 1968, despite

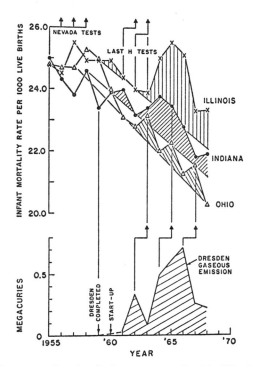

Figure 14. Infant mortality for Indiana compared with Illinois and Ohio (US Vital Statistics).

the well-known lack of sufficient medical care in the rural areas, such as those in North Dakota.

This suggests that although ordinary air pollution is undoubtedly detrimental to health, the radioactivity released by nuclear testing and nuclear plants appears to be significantly more serious in its effects on the early development of the embryo and infant.

In order to test this hypothesis further, the changes in infant mortality in the six counties immediately adjacent to the Dresden Reactor for the years following the sharpest rise in emission were compared with the changes in six control counties more than forty miles to the west. They were chosen as far away as possible in northern Illinois, not bordering either on the Illinois or Mississippi Rivers, which are known to be polluted by radioactive wastes.

The result of this test for 1966 relative to 1964 is shown in Figure 17. Again, the same general pattern is observed as for the Hanford Reactors, the nearby counties showing much greater rises than the more distant control counties.

138

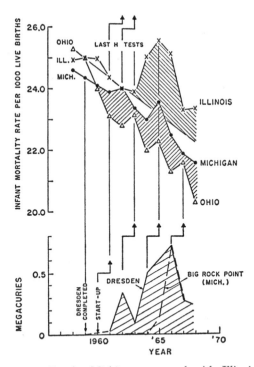

Figure 15. Infant mortality for Michigan compared with Illinois and Ohio (US Vital Statistics).

In the case of the Dresden Reactor, it is possible to carry out a still more crucial test of the biological mechanism that may be involved in bringing about such a large effect for relatively small, measured, external doses, which even in the year of peak releases did not exceed 70–80 mr at the plant boundary when the available measurements for 1967–1968 are used to calculate the dose.

As discussed briefly above and elsewhere,[21] the most serious effect is likely to be an indirect one, whereby the radiation acts on the key biochemical processes in such crucial glands controlling metabolism and growth as the pituitary and the thyroid glands. Such action could lead to a small decrease in weight at birth, or to a greater frequency of prematurity, such as has in fact been observed in animal experiments and, since the early fifties, among infants born in the USA.[22]

If this immaturity at birth leads to a reduced ability to fight off infections or to a greater likelihood of such diseases of early infancy as hyaline membrane disease, respiratory distress, and atelectasis, one would expect a higher mortality rate due to such diseases in early postnatal life.

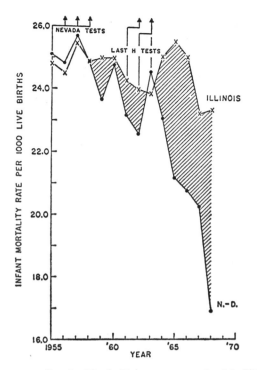

Figure 16. Infant mortality in North Dakota compared with Illinois (US Vital Statistics).

To test this hypothesis, one can compare the changes in the fraction of all births classified as 'premature' or under 2500 grams for Grundy County as compared with the changes in the control counties to the west. If immature birth is indeed the principal mechanism leading to excessive infant deaths, one would then expect to find a greater rise in the fraction of such births during the period of peak emission in Grundy than in the distant control counties.

That this is indeed the case may be seen in the plot of Figure 18. A peak in the incidence of premature births of close to 140 per cent is seen to have coincided with the peak of gaseous emission, declining again as the emissions declined, while the control counties showed no such rise. For Grundy, the increase was from 3.60 to 8.70 per cent of all births.

Thus, both radioactive releases from nuclear facilities and nuclear detonations seem to produce similar changes in the infant mortality through the indirect biochemical action of fallout on the crucial hormone-producing organs of the mother and the fetus, leading to a low-

140

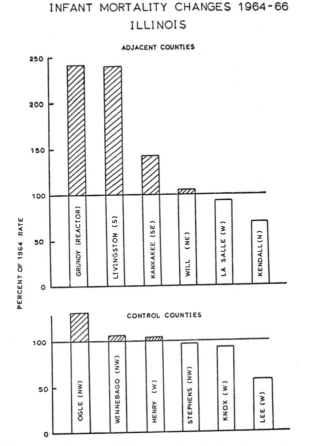

Figure 17. Percentage changes in infant mortality in the six counties surrounding the Dresden Reactor (< 30 miles distant) between 1964 and 1966 compared with the changes in six control counties to the west, following the rise in activity released from 71,600 curies in 1963 to 610,000 curies in 1965 (US Vital Statistics).

ered resistance to the environmental stress most critical shortly after birth.

Identical patterns of rises in infant mortality have now been found for two other boiling water reactors, as shown in Figure 19 for the group of small counties around the Big Rock Point Plant in Michigan, and in Figure 20 for the Humboldt Reactor near Eureka in Humboldt County, northern California. Again there is a sharp halt in the normal decline of infant mortality following release of large quantities of gaseous activity comparable to those released at the Dresden Reactor, while

141

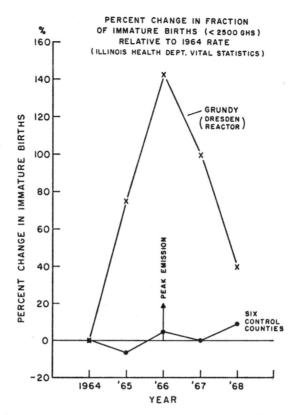

Figure 18. Percentage changes in the fraction of births under 2500 grams for Grundy County and the six control counties to the west (Illinois Vital Statistics).

more distant areas continue their decline, shown for the Humboldt area in Figure 21.

As described elsewhere in greater detail,[23] the same pattern also occurred for the commercial fuel-reprocessing plant operated by the Nuclear Fuel Services Company in West-Valley, New York, after it went into operation in April 1966.[24] In Figure 22 it is shown that in the counties of western New York, within a 30–50-mile radius, infant mortality rose sharply during the following year, whereas in the more distant counties it declined as it did in New York State as a whole. Like Humboldt County, the nearby areas had shown a peak near the height of weapons testing, then began to decline only to reverse this trend sharply after the onset of large radioactive waste releases.

A typical case is Genessee County, New York, shown in Figure 23,

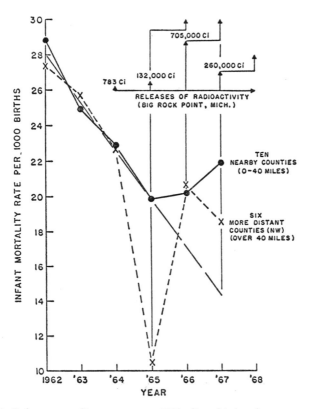

Figure 19. Infant mortality rate per 1000 live births for a group of ten counties within a radius of about 40 miles of the Big Rock Point Nuclear Plant in Charlevoix, Michigan, together with the yearly gaseous releases. The total number of deaths in these counties was 45 in 1966.

where infant mortality rates began to exceed those of the rest of the state only after onset of plant operation.

That even the relatively smaller radioactive gas releases from a gas-cooled nuclear reactor appear to be capable of producing detectable rises in infant mortality is shown for the case of the Peach Bottom Reactor located on the Susquehanna River in York County, Pennsylvania. In Figure 24 the typical drop in infant mortality after cessation of atmospheric tests for the two counties on either side of the plant, namely York and Lancaster, and the control county, Lebanon, to the north is shown. The decline continued until the onset of a large increase in emissions resulting from fuel failure that started in 1967 and reached 109 curies in 1968.[19] After 1967 York and Lancaster reversed their trend,

143

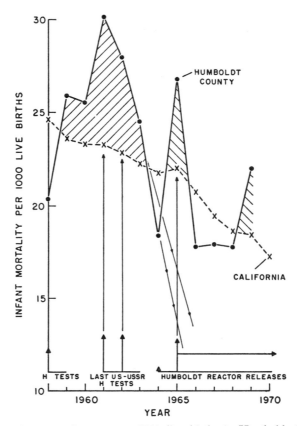

Figure 20. Infant mortality rate per 1000 live births in Humboldt County, California, 1958–1969. Releases from the Humboldt Reactor increased from 5975 curies gaseous waste in 1964 to 197,000 curies in 1965. Further rises took place in 1967–1968. Liquid waste discharges rose steadily to a peak of 3.2 curies in 1968, corresponding to 19.7 per cent of the permissible limit. Note the peaks corresponding to the 1961–1962 nuclear tests, and the steady decline of California as a whole after 1961.

and only the more distant control county 30–50 miles distant remained low.

Part of the reason why even the relatively small release from the Peach Bottom Reactor had such a strong effect seems to be that the surrounding area is a major dairy farming region, where such biologically important but relatively short-lived rare-gas daughter products as cesium-138 and strontium-89 known to be produced in large amounts from the escaping xenon-138 and krypton-89 can rapidly enter the body through the locally produced milk and other dairy products. Thus, the

144

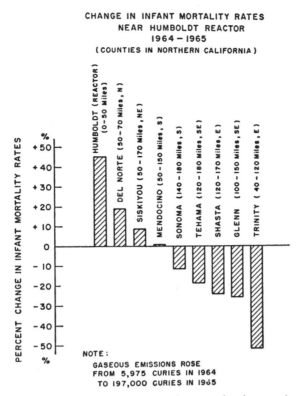

Figure 21. Percentage changes in infant mortality rates for the counties surrounding the Humboldt Reactor between 1964 and 1965, when gaseous releases rose from 5975 to 197,000 curies. Only Humboldt and Del Norte Counties immediately adjacent along the Pacific Coast showed significant rises greater than 10 per cent. All other counties either remained constant or declined, especially those separated from Humboldt by the coastal mountain ranges such as Trinity, Shasta, Tehama, and Gleen to the east and southeast.

number of curies released able to produce serious biological effects can be much smaller than from a fuel-processing plant discharging mainly krypton-85 that has no radioactive daughter product.

But the potential damage is not merely confined to the newborn and young child. There is evidence that suggests that the many radioactive gases presently released from nuclear reactors and nuclear tests may have a serious effect on the incidence of chronic diseases of the respiratory system, such as bronchitis and emphysema, that equal or even exceed the effects of conventional chemical air pollutants.

This is more strikingly shown in Figure 25, which plots the number of deaths due to respiratory diseases other than influenza and pneu-

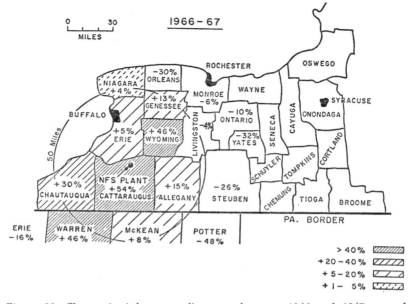

Figure 22. Change in infant mortality rates between 1966 and 1967 near the Nuclear Fuels Services plant in Cattaraugus County, New York, after it went into operation in April 1966. Note rises for ring of counties within 40–60 miles, and declines at greater distances to the east and northeast in New York. Counties in Pennsylvania along the Allegheny River flowing south from Cattaraugus County, such as Warren and Venango, also showed sharp rises in this period.

monia per 100,000 population in New Mexico and New York State between 1941 and 1965.

It is clear that between 1945 and 1950 there was a sharp rise of deaths due to noninfectious respiratory diseases such that the incidence of these diseases, previously very low in the pollution-free air of New Mexico, exceeded the death rate for the same diseases in heavily polluted New York by as much as a factor of 2.

That this is not an isolated case perhaps associated with a sudden influx of older people into New Mexico after 1950 follows from Figure 26 where similar data on deaths due to respiratory diseases have been plotted for Wyoming and Illinois. Again, there is the dramatic rise of chronic obstructive lung disease deaths in a state of almost no ordinary air pollution, such as Wyoming, to levels well above the death rates in heavily industrialized and polluted Illinois. And a similar situation exists for Wyoming relative to heavily polluted Pennsylvania, where respiratory death rates were five times higher than in Wyoming before

146

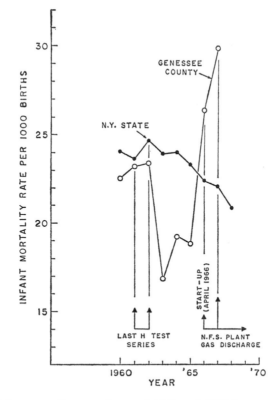

Figure 23. Infant mortality rates between 1960 and 1967 for a typical county in western New York within forty miles of the Nuclear Fuels Services Plant in Cattaraugus County. Note sharp rise above the rate for New York as a whole when plant releases started in early 1966.

nuclear testing began, while in recent years the rate in Wyoming began to exceed that in Pennsylvania (see Figure 27).

Such an apparently strong effect of radioactivity in the dry air of the west central part of the USA fits to the observed high beta-radiation activity in the dusty areas of the western states relative to that in the high rainfall areas east of the Missisippi, where the activity sinks into the soil to give lower air concentrations but higher strontium-90 levels in the milk.[25]

The operation of boiling-water nuclear reactors with their discharges of large quantities of radioactive gases appears to have had a more serious effect on the rate of non-infectious respiratory disease than the operation of fossil fuel plants (see Figure 26). In the decade 1949–1959 prior to the start of Dresden releases, the mortality rate for these dis-

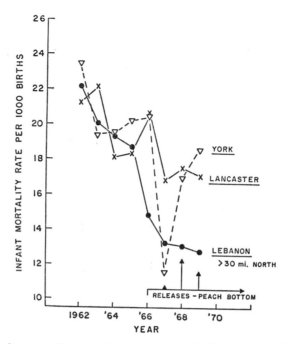

Figure 24. Infant mortality rates for the area near the Peach Bottom Reactor, York County, Pennsylvania, before and after onset of significant gaseous releases in 1967–1968, compared with rates in nearby Lancaster, directly adjacent to the east of the reactor, and Lebanon, more than thirty miles to the north of Lancaster. Releases were 0.00126 curies in 1966, 7.76 curies in 1967, 109 curies in 1968, and 100 curies in 1969.

ease rose only some 10 per cent despite a 100 per-cent increase in power generated. But in the years following onset of Dresden operations, the rate of rise increased almost ten-fold, exceeding that of either New York or Pennsylvania. And since the onset of Dresden emissions, respiratory diseases and bronchitis as a cause of death in infants older than 28 days in Illinois showed the sharpest rise among all causes of death.[26]

However, there are strong indications that gaseous radioactive discharges may not be the only source of significant effects on health. Upward changes in the steady downward trend of infant mortality have occurred in the county containing the Shippingport Pressurized Water Reactor. These are closely correlated with known peaks of tritium and other radioactivity in the liquid effluent.

As seen in Figure 28, the steady decline in infant mortality rates since the end of World War II, although slowed temporarily during the period of the Nevada tests from 1951 to 1958, ended suddenly in 1960,

148

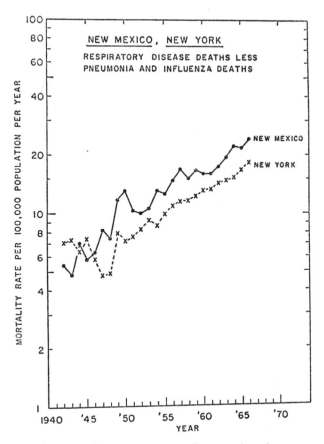

Figure 25. Mortality rate due to respiratory diseases other than pneumonia and influenza per 100,000 population for New York and New Mexico between 1941 and 1965. These diseases are principally emphysema, bronchitis, and asthma.

when both the tritium and other activity in the liquid released from Shippingport reached a sharp peak. Mortality rates then declined once more and reached a second, smaller peak when the reported beta and gamma activity rose to another high in 1964. A third sharp rise of infant mortality rates took place from a low of 16.4 per 1000 births in 1965 to a high of 24.2 by 1969, following a renewed rise in the tritium releases from 3.04 curies in 1965 to 35.2 in 1968.

Thus, by 1969, the rate of infant mortality in Beaver County, Pennsylvania, had climbed some 58 per cent above the low point reached six years earlier, in the face of a general decline of infant mortality for the USA, and Pennsylvania as a whole. And a similar reversal in the

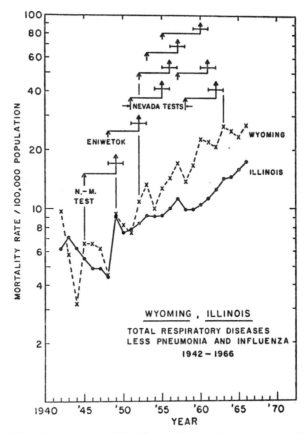

Figure 26. Mortality rate per 100,000 population for respiratory diseases other than pneumonia and influenza for Wyoming and Illinois. Note also sharp rise in Illinois after onset of Dresden operation in 1959.

downward trend took place in adjacent Columbiana County, Ohio, downstream from Shippingport, but not in upstream counties such as Allegheny County directly to the west.

It therefore appears that tritium, generally regarded as the least toxic of all isotopes, may also be more serious when incorporated into the DNA of the cells in the early developing embryo, as a number of recent animal studies have in fact also suggested.[27]

In view of the proposed large increase in the amount of nuclear generating facilities to be installed near large metropolitan areas such as New York City, it seemed desirable to carry out a study of possible health effects on children in the Greater New York metropolitan area

150

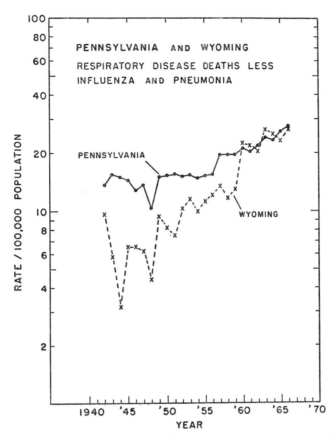

Figure 27. Mortality rate per 100,000 population for respiratory diseases other than pneumonia and influenza for Pennsylvania and Wyoming. Note the nearly constant rate for Pennsylvania between 1943 and 1956 despite a 300 per-cent increase in electric power production by fossil fuels during this period in all heavily industrialized areas of the USA.

from the releases of nuclear facilities that have been operating in this region for the past 10–15 years.

The most important sources of radioactive effluent close to the New York metropolitan area have been the Indian Point Nuclear Power Station (PWR) operated by the Consolidated Edison Company located in Westchester County along the Hudson River some twenty miles north of New York City, and the Gas-Cooled Nuclear Reactor at the Brookhaven National Laboratory of the Atomic Energy Commission near Upton, Suffolk County, Long Island. It will be shown that both of these nuclear facilities appear to have had clearly detectable effects on

151

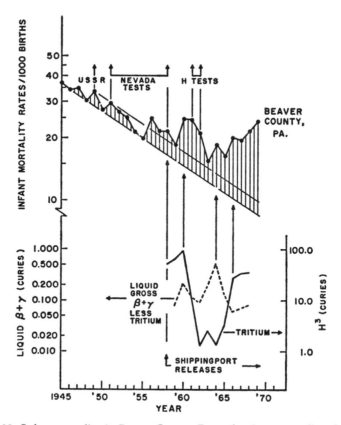

Figure 28. Infant mortality in Beaver County, Pennsylvania, surrounding the Shippingport Nuclear Reactor for the period 1945–1969. Also shown are the liquid radioactive waste releases in the form of tritium and gross beta-gamma activities that began in 1958. Note the peak in mortality rates beginning in 1960 before the 1961–1962 test series began, and the second rise associated with sharply rising tritium releases beginning in 1966.

infant mortality and leukemia rates in the surrounding counties, highly correlated with the known rises and declines of radioactive releases and the dose rates from nuclear fallout as recorded at the Brookhaven National Laboratory.

The study was based on the available data for infant mortality and cancer mortality rates for all the counties of New York State within a radius of 100 miles of New York City as published in the Annual Vital Statistics Reports of the New York State Department of Health.[28] Information on the releases from the Indian Point Unit Number 1 were obtained from a report of the US Department of Health, Education and Welfare[19] published in March of 1970, as well as official AEC sum-

152

maries of reactor releases.[29] Figures on releases of liquid wastes from the Brookhaven National Laboratory as well as on external radiation doses produced by gaseous releases and fallout were obtained from a report by A. P. Hull,[30] using the average weekly dose rates at monitoring stations at the northeastern edge of the laboratory grounds and 4.8 miles away to the north.

The basic data taken from these sources are reproduced in Tables I–VIII, reproduced in the Appendix at the end of this chapter.

In order to detect a possible effect of the releases on infant mortality it was decided to search for changes with time before and after the releases began, and also to examine the pattern of infant mortality changes with distance away from the sources of radioactive gases and liquid effluent.

Since the Brookhaven Laboratories are located well to the east of New York City (approximately 50 miles from Manhattan and some 25 miles west of Nassau County), while the Indian Point plant is located some 20 miles to the north, between Rockland and Westchester Counties, they are far enough apart to separate the effects from these two sources. This is further facilitated by the fact that the Indian Point Plant did not begin to produce significant discharges until after 1963, while the Brookhaven releases began in 1951 and declined to very small values by 1964.

In order to account for such other factors as socioeconomic, medical care, diet, drugs, pesticides, climate, air pollution, infectious diseases, fallout, and various unknown factors that might influence the changes in infant mortality besides low-level radiation from plant releases, all mortality changes in the counties near the plants were compared with neighboring counties of similar socioeconomic character having no large sources of radioactive effluent.

Thus, Westchester and Rockland may be compared most closely with Nassau County, Long Island, since it has a similar total population of close to a million, a similar suburban character, and closely similar fallout levels as well as similar socioeconomic characteristics. Likewise, Suffolk County, where Brookhaven is located, can be compared most directly with neighboring Nassau, which also had the same infant mortality rates prior to the first nuclear detonations in 1944–1945.

Furthermore, as shown in the map of lower New York State (Figure 29), it is possible to use progressively more distant counties of New York State stretching toward the northwest and north as control counties.

In order to correct for the fact that these counties further to the north have a more rural character than Westchester and Rockland, and therefore different socioeconomic situations, medical care, and air pollu-

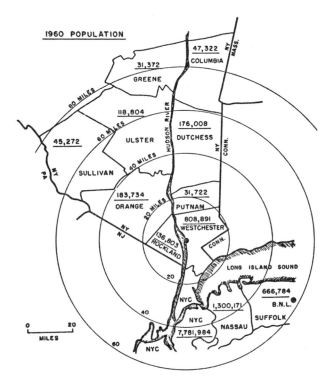

Figure 29. Map of lower New York showing the location of the Indian Point Plant in Westchester and the Brookhaven National Laboratory in Suffolk. Population figures are those for 1960.

tion, one can normalize the infant mortality rates in a suitable fashion and then examine the percentage changes following the onset of emissions. Since a given small dose of radiation is expected to have nearly the same relative effect on mortality changes regardless of the absolute rate, this technique allows one to detect changes in time as well as changes with distance from the source, despite such differences as medical care and economic level.

The counties with smaller populations can then be conveniently grouped into larger units with approximately the same distance from the point of release of the effluent.

The simplest and most direct test is to plot the pattern of mortality among infants born live and 0–1 year at death per 1000 live births for the two counties immediately surrounding the Indian Point Reactor and compare it with the time-history in Nassau County 20–50 miles away (see Figure 30).

As can be seen from an inspection of Figure 30, for a period of six

154

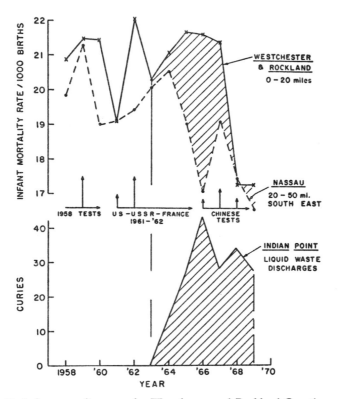

Figure 30. Infant mortality rates for Westchester and Rockland Counties compared with the rates for Nassau 1958–1969. Also shown is the liquid radioactive waste other than tritium released from the Indian Point Plant.

years prior to the onset of large releases from the Indian Point Plant in 1964, the infant mortality rates for Nassau and Westchester were essentially the same within the statistical fluctuation of about 5 per cent or ± 1.0 per 1000 births. There were rises apparently associated with the fallout from the large test series in 1958 and 1961–1962 prior to the onset of large releases from the Indian Point Plant in 1964, but the two counties showed exactly the same infant mortality rates of 19.1 in 1961, the year of lowest fallout in the air and diet just prior to the resumption of atmospheric testing by the USSR in the fall of 1961, and by the USA in 1962.

However, after the releases began from the Indian Point Reactor, while Nassau infant mortality moved downward as did most areas of the USA following the end of nuclear testing,[6,7] Westchester and Rockland moved upward and remained high for a period of four successive years. Not until after the emissions began to show a tendency to decline

155

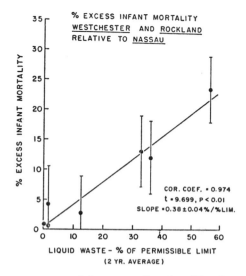

Figure 31. Percentage excess infant mortality for Westchester and Rockland Counties relative to Nassau vs. the annual amounts of liquid waste discharged from Indian Point Plant, expressed in percentage of permissible limit.

following the 1966 replacement of the original fuel-core that had developed serious leaks[19] did Westchester and Rockland infant mortality decline close to where Nassau had moved.

If one now plots the difference in infant mortality between the two counties nearest the reactor and compares it with the annual releases of liquid radioactive waste in the form of mixed fission products (beta and gamma emitters other than tritium) (see Figure 31) expressed as the percentage excess over the Nassau rate, one finds a direct linear relationship between excess mortality and the amount of activity as percentage of permissible limit.

Applying a least-square fitting procedure to the data for the period 1963–1969, one obtains a correlation coefficient $C = 0.835$. A still better fit is obtained for the two-year average, or $C = 0.974$. The t-test of statistical significance gives $t = 9.96$, which for the present case of five degrees of freedom gives a probability, P, of less than 0.01 that this correlation is a purely chance occurrence. Thus, the association between excess infant mortality near the reactor and the changing levels of liquid waste discharges must be regarded as statistically highly significant. And since, as shown in Figure 32, gaseous releases closely followed liquid releases in magnitude, not only areas bordering the Hudson River but also areas exposed to the gaseous releases would be expected to be affected.

156

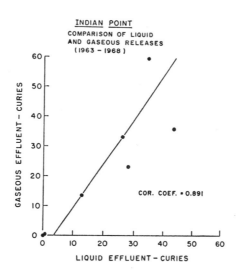

Figure 32. Correlation between liquid and gaseous effluents in the form of noble and activation gases from the Indian Point Plant 1963–1968 as reported in the P. H. S. Publication BRH-DER-70-2.

As an independent check of this result, it is of interest to compare the changes in infant mortality for the two counties near the reactor with those for counties more than forty miles to the north and north-west, namely Columbia, Greene, Sullivan, and Ulster, grouped together so as to provide a total population closer to that of Westchester and Rockland.

In order to allow such a comparison despite the more rural character of these control counties, their infant mortality rate was normalized to ⸳qual that for Westchester and Rockland in 1961, the year when Nassau showed the same infant mortality rate as the two counties next to Indian Point. In Fig. 33 the percentage changes relative to the year 1961, again both before and after the emissions began are shown.

It is seen that as in the case of the comparison with Nassau County in Figure 31, the control group shows a very similar pattern prior to 1964, but as soon as the releases occurred, a gap between the nearby and the distant counties begins to appear amounting to about four standard deviation by 1966. The control counties show a rapid decline in infant mortality while the nearby counties show a rise followed by years of failure to decline.

Once again, one can examine the correlation between the excess in the infant mortality of the exposed counties as compared to the more distant control counties, as shown in Figure 34. As in the case of the

157

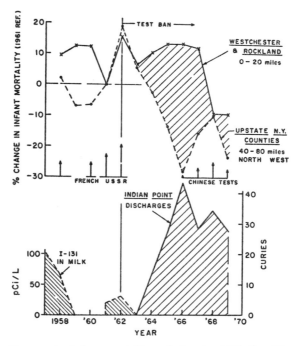

Figure 33. Changes in infant mortality relative to 1961 for Westchester and Rockland compared with four upstate control counties 40–80 miles to the north. Also shown are Indian Point liquid releases and Iodine-131 in New York City milk in average monthly concentrations (pCi/liter). Liquid release as percentage of permissible limit.

use of Nassau as a control, there is a strong, positive correlation between the excess mortality and the quantity of radioactive wastes discharged. The correlation coefficient is found to be 0.957 and $t = 7.37$, which, for the five degrees of freedom, leads again to a small probability, $P \ll 0.01$, that this association is a pure chance occurrence. Furthermore, the amount of change per unit radioactive discharge is found to be closely the same using this group of controls as when Nassau County was used, within the accuracy of the data.

Using the same normalization procedure for the group of intermediate counties to the north of Westchester and Rockland, namely Dutchess, Orange, and Putnam, it is now possible to test whether they show a pattern intermediate between the nearby and more distant counties during the period of peak emissions from the Indian Point Plant.

The result for the year of peak emission (1966) is shown in Figure 35, where the three groups of counties have been plotted according to their average distances from the Indian Point Plant in Westchester

158

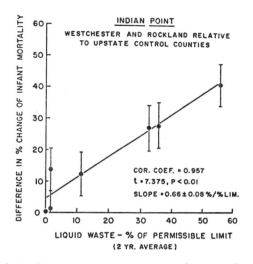

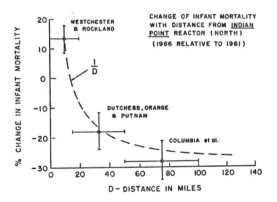

Figure 34. Correlation between percentage excess infant mortality for Westchester and Rockland Counties relative to upstate control counties and liquid waste discharges from the Indian Point Reactor.

Figure 35. Percentage changes in infant mortality by 1966 relative to 1961 for counties at increasing distances from the Indian Point Plant moving north.

County. Not only do the intermediate counties show the required intermediate position in the change of infant mortality, but the three groups show a dependence on distance consistent with an inverse first-power law expected for long-lived gases diffusing from a stack.[31]

As a further test of the hypothesis that the infant mortality changes are associated with releases from the Indian Point Plant, one can make the same plot for Nassau and Suffolk Counties to the southeast as

159

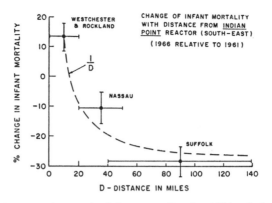

Figure 36. Percentage changes in infant mortality by 1966 relative to 1961 for counties at increasing distances from the Indian Point Plant, moving south-east.

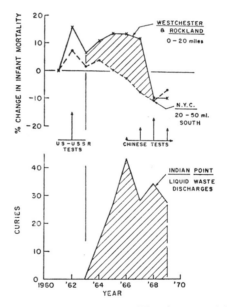

Figure 37. Changes in infant mortality for Westchester and Rockland Counties compared with New York City relative to the 1961 rates.

shown in Figure 36, and again the pattern of declining mortality fits the hypothesis.

It is of interest to see whether, despite its much poorer socioeconomic pattern, air pollution problems, and medical care, New York City

160

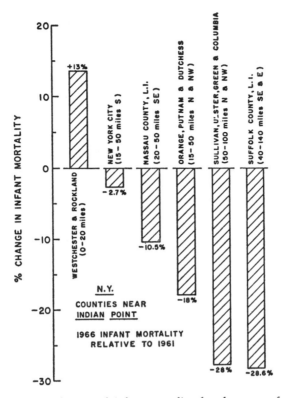

Figure 38. Percentage changes of infant mortality for the years of peak releases from the Indian Point Plant by 1966 relative to 1961 for all New York counties within a radius of 100 miles.

shows a decline in infant mortality during the time that Westchester and Rockland showed a rise above the 1961 level. Using the same normalization procedure, the infant mortality for New York City, shown in Figure 37, is in fact found to decline after 1964, though not as rapidly as the more remote counties to the north and east. Thus, the pattern of infant mortality changes following the onset of radioactivity releases from the Indian Point Plant, as shown in the bargraph of Figure 38, is consistent with a causal effect of the releases on infant mortality, similar to the effects already noted for seven other nuclear reactors and fuel processing facilities.[32]

Taking the control counties either to the north or to the east as a reference, the excess infant mortality associated with a release of 43.7 curies per year of mixed fission products in liquid waste and 36.4 curies of noble and activation gases is 41 per cent. For the year 1966, this

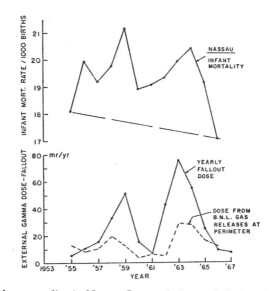

Figure 39. Infant mortality in Nassau County during period of peak nuclear testing in the atmosphere compared with external radiation levels measured at Brookhaven. Also shown are annual doses from gaseous releases measured at the northeast perimeter of the Brookhaven Laboratory.

represents an excess mortality of approximately 100 infants 0–1 year old in Westchester and Rockland Counties combined out of a total of 367 infants who died in their first year during 1966.

For New York City, assuming that the relative changes, shown for 1966 in Figure 38, can be attributed to the plant releases, the excess mortality would be approximately 26 per cent. This would mean that out of the total of 3686 infant deaths in 1966 some 750 probably died as a result of the operation of the Indian Point Plant. Thus, although New York City is more distant than Westchester and Rockland, because of its large population the total number of additional deaths is some seven times larger than that for the nearby counties.

These results are so serious that it is essential to apply still further tests in an effort to see whether the observed association is likely to be of a causal nature. Thus, if low levels of radiation near a nuclear plant, typically well below the 500 mr per year allowable to any individual, or of the order of a few millirads per year, can indeed produce such serious effects on the early embryo, then effects should be seen for the low-level fallout radiation measured at Brookhaven over a period of many years.[33]

Assuming that Nassau County on Long Island just west of Suffolk

162

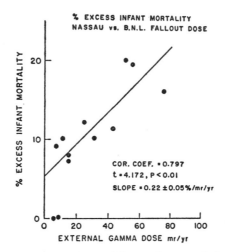

Figure 40. Excess infant mortality in Nassau County relative to the 1955–1966 base line vs. the external gamma radiation dose measured at Brookhaven National Laboratory. The slope of the least-square fitted line corresponds to a 22 per-cent increase for a dose of only 100 mr/yr.

County received essentially the same fallout levels as Brookhaven, it is possible to see whether the changing levels of annual fallout dose were in fact accompanied by corresponding changes in infant mortality in Nassau.

The data on infant mortality rates for Nassau are shown in Figure 39 for the period following the first large H-bomb tests in the Pacific in 1954, together with the annual external gamma radiation dose as measured at Brookhaven (Table VIII).

It is seen that as the radiation dose rose from about 6 mr per year in 1955 to 51.5 mr per year in 1959, infant mortality rose 17 per cent from 18.1 to 21.2 per 1000 live births. This first rise was followed by a second peak associated with the 1961–1962 test series, again followed within a year by a renewed peak in infant mortality.

Using the line connecting the point for 1955 before the rise and 1966 after the end of large-scale testing as a reference, it is possible to arrive at estimates for the yearly excess infant mortality and compare them with the measured external gamma dose.

The result of this comparison is shown in Figure 40. It is seen that the excess infant mortality in Nassau is indeed highly correlated with the changing levels of fallout radiation varying up and down as fallout levels rose and declined repeatedly. The correlation coefficient is found to be 0.797, with a t-value of 4.172, corresponding to $P < 0.01$, making it a highly significant association.

163

The slope of the line is found to be 0.22 ± 0.05 per cent mr per year. Thus, this suggests that a dose of as little as one millirad of fallout per year radiation from the ground or only about one per cent of natural background radiation leads to almost a 0.25 per-cent increase in infant mortality.

But a dose of 1 mr/yr is far below the present maximum dose of 500 mr/yr permitted by existing AEC regulations for nuclear plants. The infant mortality rises of 20—40 percent near nuclear facilities are not inconsistent with the doses that might be received, considering not only external radiation but also internal doses to critical organs of the sensitive embryo during the critical first twelve weeks of organ development, which must be added to the measured external dose.

Actually, the recent work of Stewart and Kneale[4] on the effect of diagnostic x-rays on the risk of childhood leukemia when given during intrauterine development indicates that it does lead to a comparable increase in risk. Using Stewart and Kneale's estimate of one rad to the late fetus resulting in 572 extra cases of leukemia and cancer per million population irradiated, and a normal incidence of 700 per million children born, one arrives at a doubling dose of 1200 mr. Furthermore, using Stewart and Kneale's result that the early embryo in the first trimester is some fifteen times more sensitive than the late fetus,[32] one arrives at doubling doses as low as 80 mr for the first three months of development.

Thus, an annual dose of 76 mr corresponding to the maximum observed at Brookhaven from external fallout alone might result in three-month doses as high as 25 mr to the early embryo, leading to an increase in leukemia incidence of about 30 per cent, comparable to the magnitude of the observed increases in infant mortality from all causes.

As a test of the hypothesis that such small levels of radiation can in fact lead to detectable rises in leukemia even when given over a period of months, one can examine the changes in leukemia in Nassau County.

Since the typical latency period for leukemia is some 3—5 years for the infant irradiated in utero or early postnatal life, the comparison must be carried out with the radiation level existing 3—5 years earlier.

The leukemia data for Nassau County are shown in Figure 41, together with the measured external radiation dose five years prior to the reported leukemia mortality.

Inspecting Figure 41, we find a striking parallel behavior for the two quantities; this is confirmed by the correlation plot, shown in Figure 42. The correlation between the increase in leukemia relative to 1960 and the radiation levels after 1955 is strong and positive with a correlation coefficient of 0.819, $t = 3.503$ corresponding to $P < 0.02$. The slope obtained by the least-square fit is 0.49 ± 0.13 per cent

164

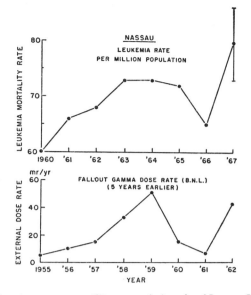

Figure 41. Leukemia rate per million population for Nassau County compared with the measured external gamma radiation rate from fallout five years earlier.

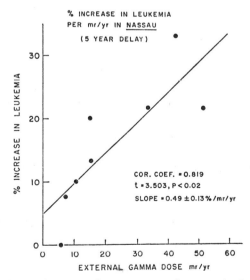

Figure 42. Correlation between the percentage increase in leukemia rates in Nassau County and the annual dose from external fallout radiation. The least-square fitted line corresponds to an increase of 49 per cent for a dose of 100 mr/yr.

mr/yr, comparable with the slope relating the percentage increase of infant mortality and fallout radiation.

From this result, one can calculate the doubling dose, or the dose for a 100 per-cent increase, of 204 ± 54 mr/yr, equal to 51 ± 13 mr in any three-month period. Considering that this represents only external dose, a total doubling dose of 80 mr to the early embryo as obtained from the study of diagnostic x-ray effects is therefore not unreasonable for fallout radiation as well.

One should therefore not be surprised to find similar changes in infant mortality that involve subtle genetic defects leading to slight immaturity at birth, which by itself tends to increase greatly the chance of death from respiratory or infectious diseases.[22] Such changes in immaturity or lowered weight at birth have in fact been observed in animal studies[27] and among children born in the USA since the early fifties,[22] the time when large-scale nuclear testing began, a trend that has only recently begun to reverse itself.

In fact, mortality for all age groups showed sharp upward changes beginning in the early fifties, as first pointed out by I. M. Moriyama.[9]

These considerations, therefore, lead one to expect that the gaseous and liquid effluents from the Brookhaven Reactor may also have led to detectable changes in infant mortality in Suffolk County.

That this appears in fact to have been the case is shown in Figure 43, where the infant mortality in Suffolk County is shown together with the reported liquid effluent produced and discharged at Brookhaven. The anomalous rise of infant mortality in Suffolk between 1953 and 1960 relative to Nassau is strongly associated with the reported activity produced at Brookhaven and the fraction released into the streams.[30] Both before and after this period, Suffolk and Nassau showed the same infant mortality rates. And with the drastic reduction in releases that took place since the peak of activity in 1959, infant mortality in Suffolk County dropped from a high of 24.1 in 1960 to an all-time low of 17.0 in 1969, an unprecedented drop of 30 per cent in only nine years.

On a number of occasions, significant quantities of radioactivity have escaped into the atmosphere from underground nuclear detonations. The most serious of these occurred at the Nevada Test Site on December 18, 1970, in the course of a test of a so-called 'tactical' weapon announced to be in the 20 kiloton range.[33]

A radioactive cloud was reported to have risen to an altitude of some 8000 feet, and to have drifted north and northeast toward Idaho and Montana.

As tabulated in Radiological Health Data and Reports,[34] the levels of gross beta activity in surface air, the total deposition of beta activity

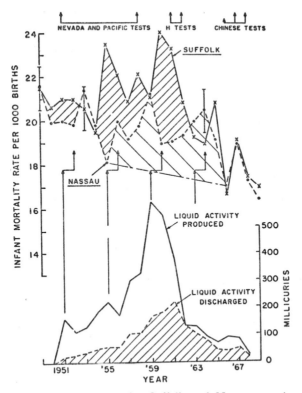

Figure 43. Infant mortality rates for Suffolk and Nassau counties, 1949–1969, compared with the releases of liquid radioactive waste from the Brookhaven National Laboratory in Suffolk County. Note that in 1949 and 1968, Nassau and Suffolk had the same rates of infant mortality. Note that throughout the period of nuclear testing and large releases from Brookhaven or from 1949 to 1962, infant mortality did not decline, and that the sharp decline began only after 1965, when dietary levels of radioactivity had sharply declined all over the United States.

on the ground, and the levels of cesium-137 in the milk increased during the month of December 1970 over large areas of the northwestern and central USA as well as southern Canada.

Upward changes in the levels of cesium-137 in milk were recorded from many states relative to the previous year's average. The highest levels were recorded in Montana, the rise being 13 pC/1, corresponding to an increase of 186 per cent. No rise or actual declines took place throughout the entire southern USA, with the exception of Texas, Tennessee, and Florida. Rises also took place across the entire northern USA and southern Canada, together with rises in air and ground activities for states more than 3000 miles from the test site.

167

The Monthly Vital Statistics Reports published by the National Center for Health Statistics of the US Department of Health, Education and Welfare for the first three months of 1971 were examined for changes in infant mortality rates in each state. In the month of January 1971 relative to the average for 1969 and 1970, infant mortality rose sharply in the states where radioactivity was deposited most heavily on the ground and in the milk, while it declined for the more distant states and the USA as a whole.

Thus, infant mortality in Idaho rose by 49.8 per cent and in Montana by 42.8 per cent, while in states that escaped the heaviest fallout, such as Louisiana, it declined 22.9 per cent. However, in Maine – far to the north and in the path of the deposition pattern that stretched across eastern Canada – a rise of 24 per cent is shown, which is presumably due to localized 'rain-outs'.

A similar pattern was found to hold for the months of January and February combined relative to the corresponding periods in 1968 and 1970. The same pattern but with lower increases persisted for the period from January to March 1971, thus confirming further the hypothesis that the rises are causally connected with the radioactive release from the December 18, 1970, test.

Thus, the most recent accidental release of radioactivity into the environment for which complete documentation of radiation levels exists confirms the very similar rises in infant mortality observed for the states downwind from the very first atomic test in New Mexico in July of 1945. In both cases, radiation levels were roughly comparable, since the 1945 Trinity test was also in the range from 10 to 20 kilotons detonated near the ground.

Thus, the testing of nuclear weapons, even underground, must be regarded as continuing to exact a heavy price in human lives through the unplanned accidental release of what have until now been regarded as harmless quantities of radioactive fission products not dangerous to human health.

Summary and conclusion

The evidence of rises in infant mortality, congenital defects and childhood cancers associated with nuclear testing has recently been corroborated by similar rises in infant mortality in the vicinity of nine different nuclear facilities known to release quantities of radioactive gases into the environment that lead to environmental activity levels comparable to those measured during nuclear weapons tests.

In both types of low-level exposure, infant mortality was associated with increased frequency of underweight birth or immaturity. Thus it

168

appears that low-level radiation acting on the early embryo fetus and young infant not only leads to significant rises in diseases previously known to be produced by radiation such as congenital defects and cancer, but also appears to act indirectly in producing small decreases in maturity at birth that in turn can increase the chance of early death from various causes such as respiratory and infectious diseases.

It therefore appears that the much greater sensitivity of the developing embryo and young infant, as compared with the adult, requires a drastic lowering of the waste discharges from nuclear facilities, as well as a complete reexamination of the balance between the biological risk to the species from all uses of nuclear fission and the benefit that is to be produced for society.

In view of the present findings, the possibility, first advanced by I. M. Moriyama, exists that both infant mortality and chronic diseases for all ages having genetic components and involving subtle disturbances of the cell chemistry may have been more seriously affected by low-level environmental radiation than had been expected on the basis of high-level radiation studies on laboratory animals carried out with external x-rays and gamma rays.

The absence of any alternative explanation of all these facts requires that, from the point of view of public health, immediate steps be taken to end all releases of nuclear isotopes into the environment from large nuclear facilities and the detonation of all nuclear devices both above and below ground.

The evidence of higher respiratory disease, childhood leukemia, and infant mortality rates near the nuclear test site in Nevada, combined with the evidence of rises and declines of infant mortality near commercial nuclear facilities following the measured levels of radioactive releases, indicates that although other air and water pollutants undoubtedly contribute to these health effects, nuclear radiation appears to be the single most serious environmental agent acting on human health on a worldwide scale at the present time.

Inasmuch as areas using conventional fossil fuels to generate electricity since the end of large-scale atmospheric testing by the US, the USSR, and the UK show lower death rates from respiratory diseases and infant mortality than the nonindustrialized areas of the mountain states in the western USA with their high levels of airborne radioactivity, the argument that fossil-fuel burning may be more deleterious to human health than nuclear power generation finds no support.

Therefore the overriding concern to protect the health and lives of the newborn as well as the adult would appear to require a worldwide moratorium on the construction of nuclear power plants until such time as extensive further statistical studies have been carried out by independent scientific groups throughout the world. Only if these studies provide a totally different explanation for the observed changes in the pattern of mortality for all ages since the onset of nuclear testing in 1945 — and one that is equally capable of explaining the facts — should a resumption of the effort to harness the fission process be permitted to continue.

In the meantime, far more intensive efforts must be made to develop alternative, non-polluting techniques of energy generation, and the entire question of the need to maintain the present rate of increase in power consumption in the technologically advanced nations must be subjected to serious examination.

In terms of needed further research in the area of radiation effects, it would appear that the following subjects require worldwide attention from the scientific community:

1. Detailed studies of the effects of the rare-earth radioisotopes on the function of the endocrine system governing the key biochemical processes, especially the control of maturation of the human fetus and the lung. These should involve laboratory studies, studies with experimental animals, and detailed, retrospective epidemiological studies of the causes of deaths in infants near areas of high radioactive fallout during past periods of radioactive releases.

2. Studies of the biological concentration mechanisms and pathways whereby relatively short-lived isotopes produced by fission could rapidly enter the local food chain and concentrate in the critical glandular system of the early embryo, fetus, and young infant.

3. Detailed studies of the action of various short- and long-lived beta-emitting isotopes on the lung tissue of the late fetus, the neonate, and the adult, leading to such possible effects as the suppression of lung surfactant formation and other degenerative changes associated with the noninfectious respiratory diseases such as emphysema, bronchitis, and asthma.

4. Large-scale epidemiological studies relating the rises of mortality rates from certain types of cancers and other chronic diseases to the changing local levels of radioactivity in the air, in the diet, and in human tissue.

5. Worldwide studies relating the patterns of infectious diseases to the patterns of radioactivity in the environment and diet, since radiation is known to affect the ability of the body to fight off infectious diseases, especially in the newborn of all species.

TABLE I. Infant mortality in New York counties adjacent to Indian Point Plant

Year	County	Births	Deaths	Rates	Relative Rates*	% Change*
1958	Rockland	2736	66			
	Westchester	15784	321			
	TOTAL	18520	387	21.0	109.9%	+ 9.9
1959	Rockland	2876	61			
	Westchester	15726	340			
	TOTAL	18602	401	21.6	113.1%	+13.1
1960	Rockland	3044	75			
	Westchester	15938	334			
	TOTAL	18982	409	21.5	112.6%	+12.6
1961	Rockland	3186	53			
	Westchester	16024	314			
	TOTAL	19210	367	19.1	100 %	0.0
1962	Rockland	3238	80			
	Westchester	15622	337			
	TOTAL	18860	417	22.1	115.7%	+15.7
1963	Rockland	3340	77			
	Westchester	15750	310			
	TOTAL	19090	387	20.3	106.3%	+ 6.3
1964	Rockland	3456	58			
	Westchester	15366	340			
	TOTAL	18822	398	21.1	110.5%	+10.5
1965	Rockland	3554	79			
	Westchester	14634	315			
	TOTAL	18188	394	21.7	113.6%	+13.6
1966	Rockland	3576	78			
	Westchester	13692	295			
	TOTAL	17268	373	21.6	113.1%	+13.1
1967	Rockland	3492	67			
	Westchester	13207	290			
	TOTAL	16699	357	21.4	112.0%	+12.0
1968	Rockland	3391	60			
	Westchester	12890	221			
	TOTAL	16281	281	17.3	90.6%	− 9.4
1969	Rockland	3625	56			
	Westchester	13292	236			
	TOTAL	16917	292	17.3	90.6%	− 9.4

TABLE II. Infant mortality in New York Counties 15—50 miles north of Indian Point Plant

Year	County	Births	Deaths	Rates	Relative Rates*	% Change*
1958	Dutchess	3674	77			
	Orange	4106	107			
	Putnam	632	19			
	TOTAL	8412	203	24.1	100%	0.0
1959	Dutchess	3736	98			
	Orange	4060	102			
	Putnam	696	17			
	TOTAL	8492	217	25.6	106.2%	+ 6.2
1960	Dutchess	3912	106			
	Orange	4066	118			
	Putnam	708	11			
	TOTAL	8686	235	27.1	112.8%	+12.8
1961	Dutchess	3912	109			
	Orange	4084	84			
	Putnam	770	18			
	TOTAL	8766	211	24.1	100%	0.0
1962	Dutchess	3998	93			
	Orange	4056	97			
	Putnam	724	24			
	TOTAL	8778	214	24.4	101.3%	+ 1.3
1963	Dutchess	4014	77			
	Orange	4176	102			
	Putnam	806	19			
	TOTAL	8996	198	22.0	91.3%	− 8.7
1964	Dutchess	4148	83			
	Orange	4244	113			
	Putnam	818	13			
	TOTAL	9210	209	22.7	94.2%	− 3.8
1965	Dutchess	3988	82			
	Orange	3978	98			
	Putnam	800	14			
	TOTAL	8766	194	22.1	91.7%	− 8.3
1966	Dutchess	3680	67			
	Orange	3680	73			
	Putnam	796	21			
	TOTAL	8156	161	19.7	81.7%	−18.3

* Relative to 1961 value.

Table II contd.

Year	County	Births	Deaths	Rates	Relative Rates*	% Change*
1967	Dutchess	3566	66			
	Orange	3693	71			
	Putnam	746	13			
	TOTAL	8005	150	18.7	77.6%	−22.4
1968	Dutchess	3580	72			
	Orange	3682	91			
	Putnam	846	14			
	TOTAL	8108	177	21.8	90.5%	− 9.5
1969	Dutchess	3702	69			
	Orange	3906	69			
	Putnam	987	22			
	TOTAL	8595	158	18.4	76.3%	−23.7

TABLE III. Infant mortality in New York counties 50—100 miles north and northwest of Indian Point Plant

Year	County	Births	Deaths	Rates	Relative Rates*	% Change*
1958	Columbia	918	14			
	Greene	568	14			
	Sullivan	886	20			
	Ulster	2632	74			
	TOTAL	5004	122	24.4	101.7%	+ 1.7
1959	Columbia	870	19			
	Greene	596	16			
	Sullivan	892	20			
	Ulster	2670	57			
	TOTAL	5028	112	22.3	92.1%	− 7.1
1960	Columbia	918	15			
	Greene	612	19			
	Sullivan	854	26			
	Ulster	2708	54			
	TOTAL	5092	114	22.4	93.3%	− 6.7
1961	Columbia	928	23			
	Greene	584	9			
	Sullivan	896	33			
	Ulster	2720	58			
	TOTAL	5128	123	24.0	100%	0.0

*Relative to 1961 value.

Table III contd.

Year	County	Births	Deaths	Rates	Relative Rates*	% Change*
1962	Columbia	924	21			
	Greene	562	14			
	Sullivan	824	31			
	Ulster	2574	74			
	TOTAL	4884	140	28.7	119.6%	+19.6
1963	Columbia	964	22			
	Greene	568	13			
	Sullivan	940	30			
	Ulster	2536	63			
	TOTAL	5008	128	25.6	106.7%	+ 6.7
1964	Columbia	892	17			
	Greene	600	16			
	Sullivan	876	26			
	Ulster	2532	56			
	TOTAL	4900	115	23.5	97.9%	− 2.1
1965	Columbia	844	12			
	Greene	518	5			
	Sullivan	852	18			
	Ulster	2468	63			
	TOTAL	4682	98	20.9	87.1%	−12.9
1966	Columbia	772	9			
	Greene	512	7			
	Sullivan	826	26			
	Ulster	2396	36			
	TOTAL	4506	78	17.3	72.1%	−27.9
1967	Columbia	792	19			
	Greene	493	4			
	Sullivan	754	22			
	Ulster	2214	41			
	TOTAL	4253	86	20.2	84.2%	−15.8
1968	Columbia	686	18			
	Greene	466	12			
	Sullivan	749	17			
	Ulster	2129	40			
	TOTAL	4030	87	21.6	90.0%	−10.0
1969	Columbia	807	11			
	Greene	490	9			
	Sullivan	802	23			
	Ulster	2254	37			
	TOTAL	4353	80	18.4	76.7%	−23.3

TABLE IV. Infant mortality in New York City 15—50 miles south of Indian Point Plant

Year	Births	Deaths	Rates	Relative Rates*	% Change*
1958	159,256	4279	26.9	104.7%	+ 4.7
1959	159,498	4273	26.8	104.3%	+ 4.3
1960	157,706	4142	26.3	102.3%	+ 2.3
1961	160,396	4119	25.7	100.0%	0.0
1962	157,908	4366	27.6	107.4%	+ 7.4
1963	160,582	4119	26.1	101.6%	+ 1.6
1964	159,206	4289	26.9	104.6%	+ 4.6
1965	152,900	3946	25.8	100.4%	+ .4
1966	147,530	3683	52.0	97.3%	− 2.7
1967	140,368	3344	23.8	92.6%	− 7.4
1968	131,457	3034	23.1	89.9%	−10.1
1969	135,732	3315	24.4	94.9%	− 5.1

* Relative to 1961 value.

TABLE V. Infant mortality in Nassau County, Long Island, New York, 20—50 miles south of Indian Point Plant

Year	Births	Deaths	Rates	Relative Rates*	% Change*
1958	26,088	516	19.8	103.7%	+ 3.7
1959	25,406	540	21.3	111.5%	+11.5
1960	25,298	480	19.0	99.5%	− .5
1961	24,544	470	19.1	100%	0.0
1962	23,674	460	19.4	101.6%	+ 1.6
1963	23,040	462	20.1	105.2%	+ 5.2
1964	22,178	457	20.6	107.9%	+ 7.9
1965	21,110	405	19.2	100.5%	+ .5
1966	19,704	336	17.J	89.5%	−10.5
1967	18,240	348	19.1	100%	0.0
1968	17,547	306	17.4	91.1%	− 8.9
1969	17,526	289	16.5	86.4%	−13.6

* Relative to 1961 value

TABLE VI. Infant mortality in Suffolk County, Long Island, New York, 40–140 miles southeast of Indian Point Plant

Year	Births	Deaths	Rates	Relative Rates*	Change*	% Change †
1958	14,522	322	22.2	94.9%	− 5.1	+13.8
1959	15,718	331	21.1	90.2%	− 9.8	+ 8.2
1960	17,068	411	24.1	103%	+ 3.0	+23.6
1961	17,906	419	23.4	100%	0.0	+16.7
1962	18,304	383	20.9	89.9%	−10.1	+ 6.7
1963	19,362	373	19.3	82.5%	−17.5	− 1.0
1964	19,860	377	19.0	81.2%	−18.8	− 2.6
1965	19,124	400	20.9	89.3%	−10.7	+ 6.7
1966	18,626	311	16.7	71.4%	−28.6	− 1.7
1967	18,510	356	19.2	82.1%	−17.9	− 1.6
1968	18,275	319	17.5	74.8%	−25.2	−11.4
1969	19,569	256	17.0	72.6%	−27.4	−12.8

* Relative to 1961 value.
† Relative to lowest rate attained previously (19.5 in 1954).

TABLE VII. Radioactive waste discharges from Indian Point Unit Number 1

Year	Gaseous waste, noble and active (Curies)	Tritium in liquid waste (Curies)	Liquid waste, gross beta and gamma (Curies)	Liquid waste, gross beta and gamma, *as % of permissible limit*	
				1 yr average	2 yr average
1963	0.0072	N.R.	0.164	0.26	0.24
1964	13.2	N.R.	13.0	22.0	11.13
1965	33.1	N.R.	26.3	43.0	32.50
1966*	36.4	125	43.7	70.1	56.50
1967	23.4	297	28.0	1.55+	35.80
1968	59.7	787	34.6	1.65+	1.60
1969	600	1100	28.0	1.50+	1.58

Source: Taken from US Public Health Service Report BRH/DER 70—2 (March 1970) (see Reference 6), and, for 1966, from AEC Report, Testimony of Commissioner J. T. Ramey (Reference 7).
* New fuel core installed.
+ Based on radionuclide analysis.
N.R. Not reported.

TABLE VIII. External background radiation dose rates and waste discharges at Brookhaven National Laboratories

Year	Total* mr/wk	Fallout mr/wk	Dose/yr* mr/yr	Dose/yr+ BNL release mr/yr	Liquid waste input to BNL filter bed mCi/yr	Liquid waste released from BNL filter bed mCi/yr
1949	1.80	0.21	10.9	—	—	—
1950	1.74	0.15	7.8	—	—	—
1951†	1.59	0.00	0.0	5.2	160.5	21.5
1952	—	0.03**	1.5**	3.6**	116.6	27.9
1953	1.73	0.14	7.3	3.1	132.9	35.8
1954	1.66	0.07	3.7	5.2	182.1	48.5
1955	1.70	0.11	5.7	13.5	223.8	75.0
1956	1.79	0.20	10.4	7.8	170.0	55.0
1957	1.89	0.30	15.6	10.4	300.8	105.1
1958	2.23	0.64	33.2	20.8	325.1	106.0
1959	2.58	0.99	51.5	6.8	586.6	169.5
1960	1.88	0.29	15.1	3.6	542.9	177.8
1961	1.73	0.14	7.3	7.3	384.4	219.1
1962	2.41	0.82	42.8	5.2	128.9	135.9
1963	3.05	1.46	76.0	29.6	127.5	99.4
1964	2.65	1.06	55.2	28.6	89.0	76.4
1965	2.07	0.48	25.0	15.6	66.8	41.8
1966	1.77	0.18	9.4	12.0	85.1	37.2
1967	1.73	0.14	7.3	4.7	81.2	47.9
1968	1.70	0.11	5.7	2.6	21.5	16.2
1969	1.65	0.06	3.1	0		

Source: Based on data by A. P. Hull (see Reference 8).
* Measured at 4.8 miles north of BNL Perimeter.
+ Difference between dose measured at Northeast Perimeter Station and station 4.8 miles north.
† Year of lowest background rate at station 4.8 miles north of BNL perimeter, taken as normal background rate prior to major weapons testing and releases from BNL.
** From measurements at station 3.5 miles south of BNL perimeter.

CITED REFERENCES AND NOTES

1. A. Stewart, J. Webb and D. Hewitt, *British Medical Journal* 1, 1958, p. 1495.
2. B. MacMahon, *J. National Cancer Inst.* 28, 1962, p. 1773.
3. E. J. Sternglass, *Science* 140, 1963, p. 1102.
4. A. Stewart and G. W. Kneale, *Lancet* 1, 1970, p. 1185.
5. R. Lapp, *Science* 137, 1962, p. 756.
6. E. J. Sternglass, 'Radiation Biology of the Fetal and Juvenile Mammal,' *AEC Symposium Series* 17, December 1969, p. 693 (Proc. 9th Hanford Biol. Symp.).
7. E. J. Sternglass, *Bulletin of the Atomic Scientists* 25, April 1969, p. 18.
8. I. M. Moriyama, *Public Health Reports* 75, No. 5, May 1960.
9. I. M. Moriyama, *National Center for Health Statistics, Monograph Series 3* No. 1, 1964.
10. E. J. Sternglass, *Quarterly Bulletin, American Association of Physicists in Medicine* 4, No. 3, September 1970.
11. United Nations Scientific Committee on the Effects of Radiation, 24th Session, *Supplement* 13 (A/7613), 1969.
12. 'Major Activities in the Atomic Energy Programs,' *Semi-Annual Reports,* January 1952 and later years.
13. C. W. Mays, *Hearings on Fallout, Radiation Standards and Countermeasures,* Joint Committee on Atomic Engergy, Part 2, August 1963, pp. 536—563; also R. C. Pendleton, R. D. Lloyd and C. W. Mays, *Science* 141, 1963, p. 640.
14. E. Reiss, *Hearings on Fallout Radiations Standards and Countermeasures,* Joint Committee on Atomic Energy, Part 2, August 1963, pp. 601—672.
15. W. A. Muller, *Nature* 214, 1967, p. 931.
16. E. Spode, *Z. Naturforschung* 13b, 1958, p. 286.
17. E. H. Graul and H. Hundeshagen, *Strahlentherapie* 106, 1958, p. 405.
18. 'Meteorology and Atomic Energy,' in *US AEC, Div. Tech., Inf.* (D. H. Slade Ed.) TTD-24190, 1968, Chapter 1, Section 2.2, pp. 5ff.
19. 'Radioactive Waste Discharges to the Enviromment from Nuclear Power Facilities,' US Dept. of HEW, PHS, Bur. Rad. Health Rockville, Md., March 1970 (BRH-DER 70-2).
20. 'Radiological Surveillance Studies at a Boiling Water Nuclear Power Station,' US Dept. HEW, PHS, Bur. Rad. Health, Rockville, Md., March 1970 (BRH-DER 70-1).
21. E. J. Sternglass, *Bulletin of the Atomic Scientists* 26, No. 5, May 1970, p. 41.
22. Helen C. Chase and Mary E. Byrnes, 'Trends in Prematurity in the United States,' *American Journal of Public Health* 60, October 1970, p. 1967.
23. E. J. Sternglass, 'Infant Mortality Changes Near a Nuclear Fuel Reprocessing Facility, University of Pittsburgh, Pa., November, 1970.
24. B. Shleien, 'An Estimate of Radiation Does Received in the Vicinity of a Nuclear Fuel Reprocessing Plant,' US Dept. HEW, Bur. Rad. Health, Rockville, Md., May 1970 (BRH-NERHL 70-1); also BRH-HERHL 70-3, July 1970.
25. *Radiological Health Data and Reports,* published monthly by Bur. Rad. Health, Rockville, Md.
26. 'Illinois vital Statistics,' Illinois Dept. of Health, Springfield, Ill., 1963—1968, Table D: Ten Leading Causes of Infant Death.
27. Y. I. Moskalev et al., 'Radiation Biology of the Fetal and Juvenile Mammal,' *AEC Symposium Series* 17, December 1969, p. 153; T. M. Fliedner et al., *ibid.,* p. 263; D. F. Chaill and C. L. Yuile, *ibid.,* p. 283.

178

28. Annual Statistical Reports, New York State Dept. of Health, Hollis S. Ingraham, Commissioner, Albany, N. Y. (available through 1967).
29. Testimony of Commissioner James T. Ramey, *Hearings before the Pennsylvania Senate Select Committee on Reactor Siting,* October 1970.
30. Andrew P. Hull, 'Background Radiation Levels at Brookhaven National Laboratory,' *Report,* submitted May 15, 1970 at the licensing hearings, Shoreham Nuclear Plant (AEC Docket No. 50—322).
31. M. J. May and I. F. Stuart, 'Comparison of Calculated and Measured Long Term Gamma Doses from a Stack Effluent of Radioactive Gases' in *Environmental Surveillance in the Vicinity of Nuclear Facilities,* New York, Charles C Thomas, 1970, p. 234.
32. Table III of Reference 4 gives sensitivity in the first trimester as $(8.25 - 1.00)/(1.48 - 1.00) = 15$ times relative to the last trimester. More recently a possible question as to the exact dose in the trimester has occurred (Stewart and Kneale, *Lancet,* December 5, 1970), but the ten-fold greater sensitivity of the early embryo has been well established in animal studies.
33. Associated Press, Mercury, Nevada, December 18, 1970.
34. *Radiation Health Data and Reports* 12, No. 4, April 1971, Environmental Protection Agency, Rockville, Md.

Against Pollution: Heat

E. L. BOURODIMOS

Thermopollution in the Aquatic Environment

Environmental problems, planning, and strategy on environmental pollution abatement and control are complex.[1] They are related to issues and questions of economy, esthetics, social and political status quo, and administration, all in a continuous flux and change of perspective.

Local, city, county, regional, and national organizational frameworks for the delegation and administration of functions to control pollution are indispensable for attacking and solving these problems. International mobilization — an 'international constitution' — is urgently needed in the battle for saving the planet Earth.

The human environment has become a new challenge for the United Nations. In the words of the Secretary General:

Like it or not, we are all travelling together on a common planet . . . We have no rational alternative but to work together to make it an environment in which we and our children can live full and peaceful lives . . . Perhaps it is the collective menaces . . . which will bind together nations, enhance peaceful cooperation and surmount, in the face of physical danger, the political obstacles to mankind's unity . . . If present trends are allowed to continue the future of life on earth could be endangered.[2]

The cost of industrial progress and growth has been put in question: rivers, lakes, and streams are heavily polluted; their biota are dead. A deeply serious biological and ecological instability is induced. Carbon monoxide and sulfur dioxide, soot and fly ash, acids and detergents, strontium-90 and sonic boom, fish rotted on polluted shores are but a few of the consequences and costs of industrial growth. There is an urgent need for restoring priorities above confusion. In the words of McGeorge Bundy:

The nation's political process will be sharply tested by any serious approach to environmental preservation and restoration ... There will be sharp political conflict over the assignment of these additional cost burdens ... We will be enormously dependent on the ability of men of scholarship and knowledge to communicate dangers and to explain the range of promising strategies and operational urgencies in terms that are understandable to the general public and to those with political responsibility for action. We will also need a political process that is both open and coherent. In the end, effective translation of the desire of man to preserve his environment will depend on the skill of the public man.[3]

Barry Commoner points out another crucial aspect of the environmental crisis in the USA:

Nearly all of the stresses that have caused the environmental breakdown — smog, detergents, insecticides, heavy use of fertilizers, radiation — began about 20 to 25 years ago. That period saw a sharp rise in the per capita production of pollutants. For example between 1946 and 1966 total utilization of fertilizers increased about 700 percent, electric power nearly 400 percent, and pesticides more than 500 percent. In that period the US population increased by only 43 percent. This means that the major factor responsible for increasing pollution in the US since 1946 is not the increased number of people, but the intensified effects of economically faulty technology on the environment.[4]

On the other hand, technology is an important part of modern culture and civilization. As Woodring has remarked, 'through the scientific discoveries of the nineteenth century and the resulting industrial revolution, man transformed his world, prolonged his life and produced new comforts and conveniences.'[5]

There is a dynamic equilibrium between technology and environmental pollution, between technological development and population, scientific achievements and world peace. Many people today enjoy shorter working hours, better working conditions, leisure, conveniences, and a host of other benefits and pleasures because of technical progress. On the other hand, however, uncontrolled technology threatens to destroy the biological and environmental balance; and with the perfection of nuclear weapons that can eliminate life on earth, 'the ultimate in biological absurdity' has been achieved.[6]

If reason prevails over international power politics, the planetary ecosystem can be saved. According to Burnet, 'there are three imperatives: to reduce war to a minimum; to stabilize human population; and to prevent the progressive destruction of the earth's irreplaceable resources.'[7] With the prevalence of reason and the rational use of science and technology — not as a curse but as mankind's hope — the children of tomorrow may not be the progeny of those who produced a monster they could not control.[8]

Thermopollution

Water is the medium of life, said Thales, founder of natural philosophy and science, two and a half millennia ago. And in 1913 Laurence Henderson elaborated on the same theme:

> Water, of its very nature, as it occurs automatically in the process of cosmic evolution, is fit, with a fitness no less marvelous and varied than that fitness of the organism which has been won by the course of organic evolution . . . The fitness of the environment results from characteristics which constitute a series of maxima — unique or nearly unique properties of water, carbonic acid, the compounds of carbon, hydrogen and oxygen and the ocean — so numerous, so varied, so nearly complete among all things which are concerned in the problem that together they form certainly the greatest possible fitness. No other environment consisting of primary constituents made up of other known elements, or lacking water and carbonic acid, could possess a like number of fit characteristics or such highly fit characteristics, or in any manner such great fitness to promote complexity, durability and active metabolism in the organic mechanism we call life . . .[9]

The aquatic environment in dynamic equilibrium with its biological communities is a real wonder. When photosynthetic processes through light energy absorption and transmission are in order, and self-purification and renewal secured, the biota community is harmoniously balanced. This is the hour of serenity. But any radical changes in this delicate balance may set off chemical, physical and/or biological instability, which in turn creates a fearful situation. Contamination by toxic material, DO (dissolved oxygen) depletion by algae or excessive waste load, increased levels of turbidity and temperature and low levels of pH, eutrophication by introduction of phosphate-nitrate nutrients may lead to lethal results. This the threshold of severe decay and death in the receiving water body, the breakdown of eco-

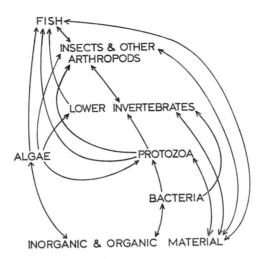

Figure 1. Simplified aquatic food web.[10]

logical balance in aquatic ecosystems. Energy and mass transfers in aquatic environments, along with a simplified aquatic food chain and web, are illustrated in Figures 1, 2, 3, 4, and 5.

Even minute amounts of toxic substances — for example cyanides, mercury, or lead — may be lethal to aquatic life. Many are not biodegradable within reasonable periods of time. Aquatic biota and marine organisms are, moreover, extremely sensitive to fluctuations in chemical, physical, and thermodynamic parameters (acidity, alkalinity, salinity, turbidity and water color, oil spills), particularly dissolved oxygen and temperature.

The most favorable pH levels for aquatic plant growth range between 7.0 and 9.2. Drops in acidity drastically influence the capability of water to support submerged aquatic plants, the basic food source for waterfowl. Similar effects are observed at high levels of bicarbonate alkalinity. Variations of 50 ppm — the result of potential changes in streamflow conditions (irrigation, flood, etc.) — can set off an unstable situation which may seriously upset the plant ecosystem.

Salinity changes may improve or deteriorate the aquatic environment, depending upon special life processes and aquatic species. High degrees of turbidity and water coloration act as serious pollutants: they may retard and diminish photosynthesis by blocking sunlight necessary for algae and aquatic vegetation. Photosynthesis is the fundamental reaction which for 'millions of years has counterbalanced death and decomposition.'[14] It is the process by which green plants and algae harness solar light energy, absorbed by chlorophyll, to build organic compounds.

186

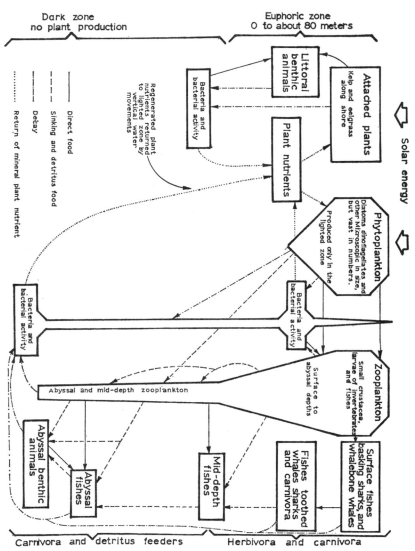

Figure 2. Major pathways of energy and material transfer in the sea.[11]

Through photosynthesis hydrogen is employed to transform carbon dioxide into carbohydrates with a simultaneous release of great quantities of free oxygen. Consequently, elimination of photosynthesis by preventing sunlight from reaching the euphotic zone results in loss of the free oxygen that is so vital to the survival of aquatic life.[15] Almost all solar energy (99.7 per cent) is absorbed and diffused in

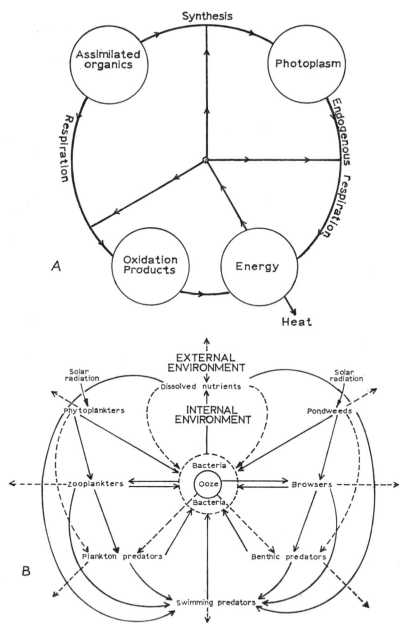

Figure 3. A: Energy in metabolism. B: Major pathways of energy and material transfer in a senescent lake, Cedar Bog Lake, Minnesota.[12]

188

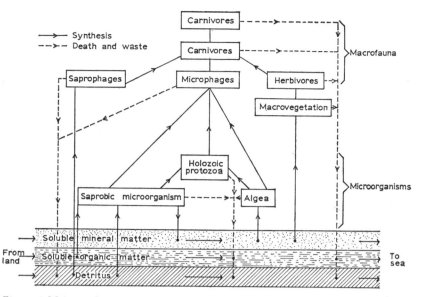

Figure 4. Major pathways of material transfer in a stream.[13]

this zone, which is also the site of the primary production of oceanic mass (phytoplankton).[16]

Oil generally removes substantial amounts of the dissolved oxygen from water environments – the oxygen required to sustain life. In Tables I, II, III, and IV surface water criteria, toxic concentrations of pesticides in fish are indicated.

Dissolved oxygen is a parameter of cardinal importance for the survival of aquatic life and functions of aquatic ecosystems as well as for the maintenance of the proper sanitation level of any body of water. Using dissolved oxygen, bacteria consume and reduce organic matter. The principal ways of replacement are through the mechanism of oxygen entrainment at the air-water interface and turbulent diffusion in the receiving water body and through uninhibited photosynthesis in the aquatic environment. The simultaneous action of deoxygenation and reæration establish the pattern of DO concentration in a stream.

This is the 'dissolved-oxygen sag' pattern and curve indicating the point of minimum DO described by Streeter and Phelps in 1925[17] and given by the equation

189

$$\frac{dD}{dt} = k'_1 L - k_2 D \tag{1}$$

where D = dissolved oxygen deficit, L = concentration of organic matter, k'_1 = coefficient of deoxygenation, and k'_2 = coefficient of reæration.

If L is described in terms of initial concentration L_0 of the organic matter in the stream (biochemical oxygen demand in mg/1) by

$$L = L_0 e^{-k'_1 t} \tag{2}$$

The solution of equation (1) using equation (2) gives

$$D = \frac{k_1 L_0}{k'_2 - k'_1} (e^{-k'_1 t} - e^{-k'_2 t}) + D_0 e^{-k'_2 t} \tag{3}$$

whereby D_0 = initial oxygen deficit at the point of waste discharge **mg/1.**

Equation (3) may be put in the following form using the relation

$$e^{-k't} = 10^{-kt}$$

whereby $k = 0.434 k'$. Hence

$$D = \frac{k_1 L_0}{k_2 - k_1} \left(10^{-k_1 t} - 10^{-k_2 t} \right) + D_0 \left(10^{-k_2 t} \right) \tag{4}$$

The proportionality factor k_1 is a function of stream temperature, whereas k_2, also a function of temperature, strongly depends upon the level of turbulence in the receiving water body. O'Connor and Dobbins give an approximate formula for the reæration coefficient of natural rivers[18] as

$$K'_2 = \frac{(D_L V)^{\frac{1}{2}}}{d^{3/2}} \tag{5}$$

where k'_2 = reæration coefficient (base e) per day, D_L = diffusivity of oxygen in water = 0.000087 ft/hr at 20 °C, V = velocity of flow (ft/hr), and d = depth of flow in ft.

The value of k'_2 ranges from 0.02 to 10.0 per day with the higher values corresponding to the steep bottom slopes of shallow streams — with high turbulence levels — and the lower values corresponding to slow-moving streams of great depths.

190

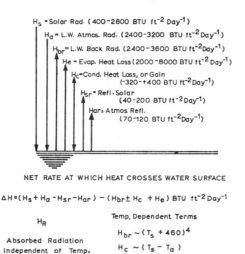

H_s = Solar Rad. (400-2800 BTU ft^{-2} Day^{-1})

H_a = L.W. Atmos. Rad. (2400-3200 BTU ft^{-2} Day^{-1})

H_{br} = L.W. Back Rad. (2400-3600 BTU ft^{-2} Day^{-1})

H_e = Evap. Heat Loss (2000-8000 BTU ft^{-2} Day^{-1})

H_c = Cond. Heat Loss, or Gain (-320-+400 BTU ft^{-2} Day^{-1})

H_{sr} = Refl. Solar (40-200 BTU ft^{-2} Day^{-1})

H_{ar} = Atmos Refl. (70-120 BTU ft^{-2} Day^{-1})

NET RATE AT WHICH HEAT CROSSES WATER SURFACE

$$\Delta H = (H_s + H_a - H_{sr} - H_{ar}) - (H_{br} \pm H_c + H_e) \text{ BTU ft}^{-2} \text{ Day}^{-1}$$

H_R

Absorbed Radiation
Independent of Temp.

Temp. Dependent Terms

$H_{br} \sim (T_s + 460)^4$

$H_c \sim (T_s - T_a)$

$H_e \sim W(e_s - e_a)$

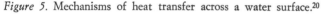

Figure 5. Mechanisms of heat transfer across a water surface.[20]

The effect of temperature on the reæration coefficient k_2 is

$$k_{2_T} = k_{2-20} \times 1.047^{T-20} \tag{6}$$

whereby k_{2_T} = reæration coefficient at the temperature T, and k_{2-20} = reæration coefficient at 20 °C.

The critical time t_c occurs when at the lower point of the dissolved-oxygen sag curve the time of change of oxygen equals zero, and the demand rate equals the reæration rate.

The dissolved-oxygen sag curve

$$k'_2 D_0 = k'_1 L = k_1 L_0 e^{-k_1 t_c}$$

Solving for t_c

$$t_c = \frac{1}{k'_2 - k'_1} \ln \frac{k'_2}{k'_1} \left[1 - D_0 \frac{k'_2 - k'_1}{k'_1 L_0} \right] \tag{7}$$

The following remarks are pertinent to equation (7): k'_1 is evaluated by BOD tests (it reflects the rate of bacterial oxygen demand); k_2 is the reæration parameter characteristic of a stream and varies at different locations of the same stream; k_1 is a constant that can be evaluated in laboratory tests, whereas k_2 should be determined from field data. It should also be stressed that the above reflect idealized analyti-

191

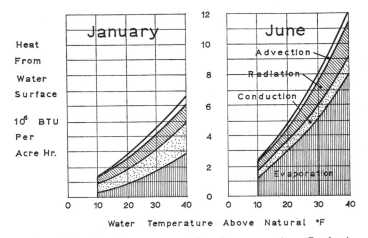

Figure 6. Heat dissipation from water surface by evaporation. Conduction and adversion during January and June.[21]

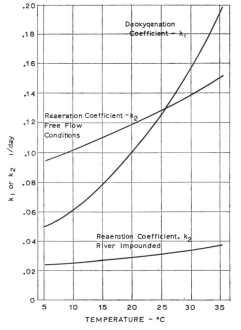

Figure 7. Self-purification as a function of temperature.[22]

192

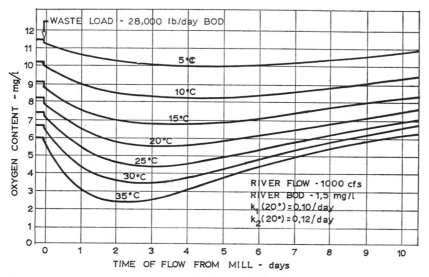

Figure 8. Oxygen sag curve — free-flowing condition.[23]

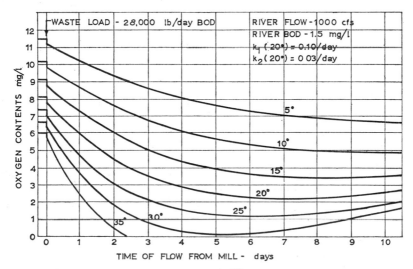

Figure 9. Oxygen sag curve — river impounded.[24]

cal assumptions: namely, it is assumed that both k_1 and k_2 are constant and that only one source of pollution exists at the particular location.[19]

Figures 5, 6, 7, 8, and 9 show the mechanism of heat transfer and dissipation across the water surface, the variation of the reæration coefficients with temperature changes, and the dissolved oxygen demand curve patterns with temperature changes. In Figure 7 the deoxygenation rate, k_1, is shown to rise markedly with increase in temperature, while the reæration rate, k_2, manifests a lower increase than k_1 for the same temperature changes. In Figures 8 and 9 the oxygen sag curve variations are illustrated.

According to Parker and Krenkel, 'The overall effects of the impoundment on the rate of oxygen recovery is demonstrated by the lower curve, which depicts the reæration rate constant under existing impounded conditions. Note that, while k_1 at a given temperature is unchanged, the value of k_2 at any temperature is significantly reduced.'[25]

Because of the water-resources development and observed temperature increases, the river, illustrated in Figure 9, can no longer assimilate the 28,000 pound BOD load under existing free conditions.

Orlob's mathematical model for simulating heat distribution in streams and reservoirs was tested on Fontana Lake with good results.[26, 27] The conclusions are pertinent:

1. The transport of heat into deep stratified impoundments is accomplished by four primary mechanisms: advection, direct solar insolation, convective mixing associated with cooling at the surface, and 'effective diffusion' identified with momentum transfer within the water body.

2. The transfer of heat energy from the epilimnion to the hypolimnion by diffusive mechanisms is limited at the thermocline by a strong density gradient.

3. A one-dimensional mathematical representation of heat energy transfer processes provides a satisfactory means of characterizing the thermal behavior of deep impoundments of the low volume-high discharge type.

4. Reservoir thermal behavior over annual diurnal cycles can be simulated with a mathematical model, using as input information continuous records of hydrologic, meteorologic, and climatologic conditions of the system.[28]

Different models of lateral and vertical heat diffusion have been developed and simulated in laboratory studies, since analytical models are generally not complete. There are different models for temperature distributions, eddy dispersion, and thermal stratification such as graphical, statistical, and semi-empirical. It has been suggested that any theoretical approach should consider internal radiation absorption, without which values of diffusivity are not representative of the physics of the phenomenon. The following remarks indicate the biology of aquatic communities and their ecological interactions with heat disposal.

1. Temperature changes in the aquatic environment are significant processes upon which aquatic life depends. Mammals generally have developed evolutionary mechanisms that keep their body temperature constant, while other animals and most aquatic organisms follow that of the environment and adapt, within limitations, to temperature changes in the environment. Survival or extinction of most species depends on this ability.

2. Thermal pollution in streams, lakes, or oceans may reach such levels that the adaptive capacity of organisms is far below that of the environment. Lower oxygen solubility due to higher temperatures combined with heavy waste load and bacteria respiration leads to an aquatic regime of low oxygen level, uncontrolled metabolic activity of thermophilic — and psychophilic organisms as well — with lethal results to biota of aquatic ecosystems. It should, however, be stressed at this point that all living organisms display a kind of internal functional independence with respect to their environment. These are the homeostatic responses, the wonder of nature, the road to survival.

3. Cellular functions nevertheless proceed only within rather short and narrow physio-chemical and biological-molecular limits within which dissolved oxygen, pH, alkalinity, and above all temperature must vary.[29]

4. Homeostasis is precisely that function of the organism which maintains an 'internal regime' within limited biochemical and physical levels, so that survival may be secured even with an increase in environmental stresses such as higher temperature or waste. Of course we neither know exactly nor in complete detail the entire spectrum of the complex biophysical, chemical, and enzymatic processes and mechanisms triggered in the course of homeostatis.

At the Portland Seminar on Thermal Pollution, Strickland (1968) observed pessimistically that 'really, the business is in such a state that we can never be certain. We don't really know to what extent the entire food chain and its indirections and feedbacks are affecting any particular part of it. The more we get to know about it, the more complicated and interacting it becomes . . .'[30]

Nonetheless, the basic patterns of the adaptability of organisms to general environmental conditions depending upon their genotype (genetic individuality) and their phenotype (phenotypic plasticity) should be considered.

5. *Acclimation,* a homeostatic mechanism, makes propagation and survival of the species possible under extreme environmental conditions which would be unsuitable without adaptive processes of the living organisms. Metabolic and enzymatic changes along with dissolved oxygen consumption — greatly influenced by temperature increases — are fundamental steps in these complex biological functions.

Thermopollution abatement techniques

There are basically three practical engineering techniques for thermopollution abatement and heat dissipation in the aquatic environment. Heated effluents from power plants may be diluted, thereby dissipating heat and reducing unwanted temperature levels. These three techniques are: 1) surface, and especially submerged, heat disposal into large bodies of receiving waters employing techniques of submerged jets, 2) cooling ponds, and 3) cooling towers.

Submerged heated effluents. The detailed hydrodynamic aspects of viscous flows,[31] along with the chemical thermodynamics in natural water 'open' systems — especially their fundamental difference in chemical reaction and equilibria as compared with idealized 'closed' isothermal systems in the laboratory[32] — and beyond that the biophysics, biology, and ecology of biota and food chain production in these environments, are challenging questions for a host of theoretical and applied disciplines.

Parker and Krenkel summarized the conclusions reached in an advanced seminar at Johns Hopkins University on mechanisms of heat energy transport and decay in natural aquatic bodies as follows:

1. The quantitative determination of heat transfer within a natural body of water, or between a body of water and its surroundings, is extremely difficult and complex.
2. Temperature change within a body of water is primarily effected by the operation, either independently or jointly, of two mechanisms: turbulent mixing of two or more batches of water of different temperatures, eventually resulting in a single batch of water at the weighed mean temperature; and heat exchange of the water with its surroundings, primarily through conduction, convection, evaporation, condensation and radiation.
3. Temperature change is affected only to a minor degree by molecular diffusion and conduction within the body of water.
4. Forecasting of heat loss from artificially heated batches of water may be attempted by use of any one or more of the three principal techniques corresponding to the main mechanisms of heat dis-

sipation. The choice will depend upon the availability of data and adherence of the specific situation to certain specialized requirements.

5. Where heat is discharged into a stream from a point source, and where complete vertical mixing may be assumed, turbulent mixing may be considered the dominant heat transfer mechanism until horizontal mixing is complete; that is, until that downstream transect is reached where the cross-section temperature is uniform from bank to bank.

6. Assuming that there is no horizontal temperature gradient, either as a result of effective turbulent mixing as described in no. 5 above, or as a result of heat discharge from a thoroughly diffused source, and further assuming complete vertical mixing, heat loss may be evaluated by either use of the heat budget method or equilibrium temperature method.

7. The complexity of the heat budget method makes this method unfeasible for field use, except where extremely precise, reliable, and rugged instrumentation is available.

8. At present, not all the instrumentation required by no. 7 above is available.

9. The relative simplicity of the equilibrium method, coupled with its feasibility for use under field conditions, suggests its use for forecasting heat loss in specific field situations using combinations of ambient conditions selected according to statistical considerations.[33]

To the above general conclusions, the following specific remarks add another serious complexity to the hydrodynamic aspect of heat disposal, particularly by free jets. Water jets issuing into a receiving water body of different density, viscosity, and temperature are mechanisms of turbulent transport in which momentum, mass, and energy (heat) are involved. This transport mechanism initiates an interplay of turbulent convection and dispersion, molecular diffusion, and gravitational convection, and finally circulation due to density gradients. Dilution of the heated efflux of submerged jets occurs through turbulent heat convection in the first phase ('forced plume') and by diffusion and gravitational convection ('plume') in a later stage of the process. An equilibrium toward lower temperature levels by lateral dispersion in the receiving water body is the final phase of the process.[34]

There are two fundamental approaches in turbulent jet flows: the microscopic-analytical and the macroscopic-integral.[35] In the microscopic approach, an evaluation of the characteristics of turbulence — being generally non-homogeneous and anisotropic — is attempted by means

197

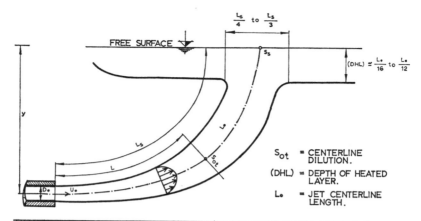

$\frac{L_s}{4}$ to $\frac{L_s}{3}$

FREE SURFACE

$(DHL) \simeq \frac{L_o}{16}$ to $\frac{L_o}{12}$

S_{ot} = CENTERLINE DILUTION.

(DHL) = DEPTH OF HEATED LAYER.

L_o = JET CENTERLINE LENGTH.

Figure 10. Definition sketch for horizontally discharged heated submerged jet.

of statistical techniques, i.e. time and space correlation analysis and spectral theory. Against this complex statistical approach Von Karman suggested the macroscopic 'eddy viscosity' concept to describe turbulent momentum transfer.[36] In order for this approach to be complete, the differential equations of heat (energy), mass, and momentum in jet flows should be formulated and solved simultaneously with proper initial and boundary conditions. Even neglecting molecular diffusion of mass and heat (generally small in magnitude compared to turbulent flows), such a formulation is very complex. The basic difficulty arises in the determination of turbulent diffusion coefficients and eddy heat inductivities. Experimental laboratory research and actual field measurements have been extensively employed for evaluation of these important coefficients in specific cases and flow fields.

Morton, Taylor, and Turner employed the integral method, and, under given assumptions, they investigated the case of gravitational convection in a rising plume from a 'point source' in a stably stratified fluid with a linear density gradient.[37] Experimental work along the same general lines but with different specific details has been performed by Morton, Abraham, Hart, Cederwall, Fan, and Forstall and Shapiro.[38] Rawn, Bowerman, and Brooks have conducted actual field investigation along with experimental work using the concepts of entrainment, dilution, and densimetric Froude number for submerged free jets.[39] Figures 10, 11, 12, and 13 show a horizontally heated submerged jet, the entrainment and velocity concepts with the similarity and trend displayed during simultaneous momentum-heat as well as momentum-mass transfer.

198

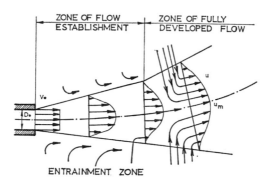

Figure 11. Development of similar velocity profiles.

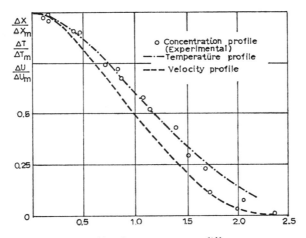

Figure 12. Dimensionless profiles for temperature difference, concentration of admixture and velocity in main region of plane jet from data given by Abramovich and Borodachev ($\bar{x} = x/b_0$).[31]

The conservation principles, the equation of state, and geometrical relationships were recently employed in a system of eleven equations — generally following the integral method — to solve the heated jet discharge in the actual case of a nuclear power plant in New York State. This jet solution approach with the accompanying heated effluent dilution complied with the very strict thermal standards of New York State. Results of previous experimental and actual field studies compared favorably with the numerical solution.[42]

The free surface heated discharge disposal is another technique, definitely less expensive than that of submerged jet discharges but possibly

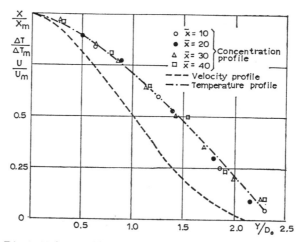

Figure 13. Dimensionless profiles for concentration of admixture by weight (helium) in main region of axially symmetric air jet in coflowing streams of air according to Forstall and Shapiro.[41]

more hazardous in terms of environmental pollution. The heated surface layer through this disposal is lighter than the cool water body beneath it. The established hydrodynamic stability between layers may be difficult to unsettle, thereby blocking circulation of currents while the surface biota and biological balance of the ecosystem may be destroyed.

The *cooling pond* is the simplest technique of cooling thermal effluents before final disposal without harm to the environment. The efficiency is limited since there is no water of air convection, but mere radiation and heat dissipation at the air-water interface.

Among the advantages are: 1) low construction cost, 2) large settling basin, and 3) benefits for other purposes, e.g. reæration.

The main disadvantages are: 1) low heat transfer rate and the necessity of large land areas, which may be costly, and 2) possible fogging — thereby adding some kind of 'vapor pollution' to the atmosphere — icing of nearby roads and potential influence upon other meteorological or climatic conditions.[43]

The *cooling tower* — mechanical or natural draft — is another cooling device which has been employed. A tower may be either 'wet' or 'dry' depending on whether it is directly exposed to the air, 'natural' or 'mechanical draft' depending on whether fans are employed for air movement, 'cross' or 'counter-flow' depending upon horizontal or vertical air flow through the heat transfer section of the tower. In mechanical draft towers, air flow can be either forced, i.e. pushed through by a fan at the base, or induced. i.e. pulled through by a fan on top.

200

Some of the advantages of evaporative mechanical cooling towers described by McKelvey and Brooke are: 1) close control of cold water temperature, 2) generally low pumping head, 3) location is not a critical factor, 4) more packing per unit volume of water, 5) a smaller approach and greater cooling range, 6) small land area requirements, and 7) less capital outlay than for a natural draft chimney. Disadvantages include: 1) high operating costs (including power), 2) high maintenance costs (both in time and money), 3) recirculation of hot humid exhaust into the air intakes, and 4) climatic variations that can affect performance because fans move a fixed volume of air regardless of its density and related heat transfer properties, efficiency decreases as wind speed increases, up to a critical velocity (at velocities above the critical value, the opposite is true because of less recirculation).[44]

Side effects may be potential fogging conditions, condensation resulting in ground-level fog or drizzle, undesirable meteorological effects. Water circulating through a condenser-cooling tower system may inhibit biological growth and metabolism, both potentially detrimental effects to aquatic environments.

The question of cooling towers versus cooling ponds or lakes for recycling at a given location or plant size depends on cost considerations and environmental variables. The question is reduced to that of determining the relative magnitude of marginal generation costs of recycling in comparison with once-through cooling.

It should be pointed out, finally, that studies examining thermal discharges and ecological unbalances are limited compared to research dealing with heat disposals to receiving water bodies. Cheney et al. summarize as follows:

Plankton is a basic unit to the food chain and depending on the type, may be nearly completely destroyed or may not be affected by thermal shock. In some thermal effluent areas, algae may thrive, occasionally to the point of being a nuisance. Invertebrate animals offer some hope, for the term 'thermal enrichment' of oysters, lobster, shrimp, etc. production can be improved by using waste to control the water temperature.

Fish are affected to various degrees by thermal additions, sometimes beneficially and sometimes detrimentally. Several sub-lethal effects of elevated water temperature are very important to consider when evaluating the extent of impact on fish from hot water discharges. The data that are available provide guidelines as to what the critical temperatures of a water body are with respect to the existing biota. The generalizations that are made can be applied to similar water bodies, but thus far the temperature problems must

201

be determined for each site, where waste heat is being produced and discharged . . ."[45]

Last but not least, a beneficial use of heat waste in non-aquatic environments may be examined. Other solutions such as commercial heating as well as industrial and household uses — subject to comparative cost (distribution, etc.) versus potential revenues — may be a means of transforming a private and public liability into a profit-earning output.

Power generation and water quality in the aquatic environment: Management, prospective cost, alternatives

Projections of power generation in the USA, which has roughly doubled every decade of this century, indicate requirement of 3640 billion KWH by 1985. This amazing trend can be seen in Table V. In Table VI the use of cooling water by various industries, with electric power utilizing over four-fifths of the total, is shown.[46]

The following data may be pertinent in this context:

1. At present thermal power generation provides 81 percent of the electricity generated in the USA and may reach 92 percent by 2000, when two-thirds of the plants will be operating on nuclear fuel.
2. Power generated by hydroelectric stations is limited, so power requirements must be increasingly met through thermal power production requiring cooling water for the plant condensers.
3. The Federal Water Pollution Control Administration estimated the cost of cooling facilities for the period 1968–1972 at $1.8 billion. By comparison, in 1965 alone about $3 billion were spent on sport fishing! Therefore, the potential effect of thermal pollution on fishing and other recreational and beneficial water uses should be considered.[47]
4. Heat waste disposal from thermal power plants (nuclear and fossil) will increase nine-fold by 2000 compared to 1968.

With these data in mind, the serious problem of power generation management assumes paramount importance: consideration of growth versus ecological, social, cultural, esthetic, and financial aspects and values. Costs and environmental reasons may dictate a rate of growth which cannot be surpassed due to environmental and social losses. Rationing and other restrictions in power consumption may be dictated by similar considerations. Such developments encroach upon private industrial interests and various aspects of public policy and administration with which we are not prepared to cope.

A systems-analysis approach might be necessary in most case studies. There are a bewildering number of subsystems in the environmental system, and their contamination and control are only remotely inter-dependent.

The pollution problems of an electric power plant burning fuel oil are different from those of a plant burning high-sulfur coal, which in turn are different from those of a plant burning low-sulfur coal. The nutrient whose availability governs the proliferation of algae in one lake may be different from the governing nutrient of another lake. Environmental problems, in short, are rarely amenable to sweeping solutions: the benefits of even major breakthroughs in research are more likely than not to be limited to discrete subsystems of the overall system.[48]

Last but not least, the method of power production by non-pollutant methods in terms of waste and thermal disposal may be considered: hydropower development and tidal power generation, geothermal and solar energy utilization. Other modern techniques might include fuel cells and magnetohydrodynamics (MHD), which would convert heat energy directly into electrical energy. MHD generators utilize the principle of passing a conductor through a magnetic field to produce current. With respect to conservation of energy resources, development of high-neutron-economy reactors, including efficient and safe types of breeder reactors, should be accelerated.

The approaching depletion of fossil fuels has made the need to conserve them for other purposes a global imperative; they are needed for petrochemicals, synthetic polymers, and other industrial and scientific uses. Substitutes are as yet unknown. Furthermore, without utilization of uranium-238 and thorium-232 through breeding reactors, the supply of uranium-235 from high-grade ores would become economically prohibitive within a few decades.

Acknowledgments

The author would like to thank Dr. J. Wiesenfeld, Chairman of the Civil and Environmental Engineering Department, and Dr. E. C. Easton, Dean of the College of Engineering, Rutgers, for their financial support and encouragement in preparing this paper.

The author expresses special thanks to Mrs. V. Georgalis-Mantzour-anis for her valuable assistance in correcting and typing the final draft of the paper.

To architect-designer H. D. Mantzouranis, who skillfully prepared the figures and drawings of this report, special appreciation is acknowl-edged.

TABLE I. Surface water criteria for public water supplies

Condition	Permissible limit	Desirable
Physical		
Color (color units)	75	Under 10
Odor	Minor	None
Temperature†		
Turbidity†		
Microbiological		
Coliforms	10,000/100/ml	Under 100/100/ml
Fecal coliforms	2,000/100/ml	Under 20/100/ml
Inorganic Chemicals (ppm)		
Arsenic	0.05	None
Barium	1.0	None
Boron	1.0	None
Cadmium	0.01	None
Chloride	250.0	Under 25
Iron	0.3	V. A.*
Lead	0.05	None
Manganese	0.05	None
Nitrates and nitrites	10.0 (as N)	None
Zinc	5.0	None
Inorganic Chemicals (ppm)		
Cyanide	0.20	None
Detergent residue (Methylene Blue active substances)	0.5	None
Oil and grease	V. A.*	None
Pesticides:		
Aldrin	0.017	None
Chlordane	0.003	None
DDT	0.042	None
Endrin	0.001	None

Source: These data summarize types of pollutants and limits cited in *Water Quality Criteria*, US Federal Water Pollution Control Administration (FWPCA).

* Virtually absent

† Temperature and turbidity levels depend upon a given aquatic environment as well as specific conditions.

TABLE II. Toxic concentrations of pesticides in fish

Material	Toxic concentration (ppm)	
Aldrin	0.02	
Chlordane	1.0	(Sunfish)
Dieldrin	0.025	(Trout)
Dipterex	50.0	
Endrin	0.003	(Bass)
Ferban fermate	1.0 to 4.0	
Methorxychlor	0.2	(Bass)
Parathion	2.0	(Goldfish)
Penta chlorophenol	0.35	(Bluegill)
Pyrethrum (Allethrin)	2.0 to 10.0	
Silvex	5.0	
Toxaphene	0.1	(Bass)

Source: FWPCA, *Water Quality Analysis,* p. 131.

TABLE III. Normal water consumption of livestock: Biological concentration potential of various domestic animals

Animal	Water consumed per day (gallons per head)
Beef cattle	7–12
Dairy cattle	10–16
Horses	8–12
Pigs	3–5
Sheep and goats	1–4
Chickens (per 100 birds)	8–10
Turkeys (per 100 birds)	10–15

Source: FWPCA, *Water Quality Criteria,* p. 130.

TABLE IV. Trace element tolerances for irrigation waters*

Element	Continuous use (on all soils)	(On fine soils) ppm / ppm / Short-term use
Aluminium	1.0	20.0
Arsenic	1.0	10.0
Beryllium	0.5	1.0
Boron	0.75	2.0
Cadmium	0.005	0.05
	5.0	20.0
Cobalt	0.2	10.0
Copper	0.2	5.0
Fluoride	No tolerance limits, may be toxic	
Iron	No problems noted	
Lead	0.05	20.0
Lithium	5.0	5.0
Manganese	5.0	20.0
Molybdenum	2.0	0.05
Nickel	0.005	2.0
Selenium	0.5	0.05
Tin	No problems noted	
Tungsten	No problems noted	
Vanadium	10.0	10.0
Zinc	5.0	10.0

Source: FWPCA, *Water Quality Criteria*, pp. 152–154.
* Concentrations greater than those indicated in this Table are considered to be agricultural toxic pollutants.

TABLE V. US electric power use

Year	Billion KWH
1912	11.6
1960	753.0
1965	1060.0
1970	1503.0
1975	2022.0
1980	2754.0
1985	3639.0

TABLE VI. Use of cooling water by US industry, 1964

Industry	Cooling water billion gallons	%
Electric power	40,680	81.3
Primary metals	3 387	6.8
Chemicals and allied products	3 120	6.2
Petroleum and allied products	1 212	2.4
Paper and allied products	607	1.2
Food and allied products	392	0.80
Machinery	164	0.30
Rubber and plastics	128	0.28
Transportation equipment	102	0.22
All other	273	0.50
Total	50,065	100.00

CITED REFERENCES

1. 'Industrial Research – News of the $27 Billion Research Industry,' *Environment: Few Solutions to Pollution,* Wescon Conference, October 1970, p. 23.
2. U-Thant, *The Human Environment – New Challenge for the United Nations,* U.N. Office of Public Information, February 1971.
3. McGeorge Bundy, 'Charting the Complexities,' *Saturday Review – SR/Environment,* April 4, 1970.
4. B. Commoner, 'Beyond the Teach-In,' *Saturday Review – SR/Environment,* April 4, 1970.
5. P. Woodring, 'Will Society Become an Uncontrollable Monster?' *Saturday Review – SR/Environment,* April 4, 1970.
6. *Ibid.*
7. Sir Macfarlane Burnet, 'Ecology and Appreciation of Life,' *The Boyer Lectures,* Australian Broadcasting Comm., Sydney, Ambassador Press, 1966.
8. H. J. Muller, *The Children of Frankenstein: A Primer on Modern Technology and Human Values,* Indiana University Press, 1969.
9. L. J. Henderson, *The Fitness of the Environment,* New York, Macmillan, 1913.
10. F. L. Paker and P. A. Krenkel, 'Thermal Pollution: Status of the Art,' *Report No. 3.,* Department of Environmental and Water Resources Engineering, Vanderbilt University, Nashville, Tenn., December 1969.
11. 'Industrial Research,' *op.cit* (see Reference 1).
12. R. L. Lindeman, 'Seasonal Food Cycle Dynamics in a Senescent Lake,' *American Midland Naturalist* 26, 1941, pp. 636–673.
13. H. A. Hawkes, 'An Ecological Approach to Some Bacteria Bed Problems (2),' *Journal of the Institute of Sewage Purification,* 1961, pp. 105–132.

14. *Encyclopaedia Britannica*, 1965, Vol. 17, p. 854.
15. C. E. Warren, in collaboration with Doudoroff, *Biology and Water Pollution Control*, Philadelphia, W. B. Saunders, 1971.
16. Guy-Harold Smith (Ed.), *Conservation of Natural Resources*, New York, John Wiley, 1971.
17. N. W. Streeter and E. B. Phelps, *U.S. Public Health Service Bulletin 146*, 1925.
18. J. D. O'Connor and W. E. Dobbins, 'Mechanism of Reaeration in Natural Streams,' *Trans-American Society of Civil Engineers* 123, 1958 p. 655.
19. J. W. Clark and W. Viessmann Jr., *Water Supply and Pollution Control*, Scranton, Pa., International Textbook Co., Second Printing, March 1966.
20. F. L. Parker and P. A. Krenkel, *op. cit.* (see Reference 10).
21. *Ibid.*
22. *Ibid.*
23. *Ibid.*
24. *Ibid.*
25. F. L. Parker and P. A. Krenkel, *Physical and Engineering Aspects of Thermal Pollution*, Cleveland, Ohio CRC Press, A Division of the Chemical Rubber Co.
26. G. T. Orlob, 'Prediction of Thermal Energy Distribution in Streams and Reservoirs,' *Report to the California Department of Fish and Game*, June 1967.
27. G. T. and Selna Orlob, 'Prediction of Thermal Energy Distribution in Deep Reservoirs,' Proceedings, *Sixth Annual Sanitary and Water Resources Engineering Conference*, Vanderbilt University, Department of Sanitary and Water Resources Engineering, Technical Report No. 13, 1967.
28. F. L. Parker and P. A. Krenkel, *op.cit.* (see Reference 10).
29. C. E. Warren, *op.cit.* (see Reference 15).
30. F. L. Parker and P. A. Krenkel, *op.cit.* (see Reference 25).
31. E. L. Bourodimos, 'Turbulent Transfer and Mixing of Submerged Heated Water Jet,' American Geophysical Union, *Journal of Water Resources Research*, April 1971.
32. J. J. Morgan, 'Applications and Limitations of Chemical Thermo-dynamics in Natural Water Systems,' *Equilibrium Concepts in Natural Water Systems*, Advances in Chemistry Series, Pittsburg, Academic Press, March 23–24, 1966.
33. F. L. Parker and P. A. Krenkel, *op. cit.* (see Reference 25).
34. E. L. Bourodimos, *op. cit.* (see Reference 31).
35. *Ibid.*
36. J. O. Hinze, *Turbulence*, Chapter 6: 'Nonisotropic Free Turbulence,' New York, McGraw-Hill, 1959; G. I. Taylor, 'Statistical Theory of Turbulence,' Parts I–V, *Proceedings, London Mathematical Society*, 20A: 196, 1935; *Idem.*, 'The Statistical Theory of Isotropic Turbulence,' *Journal of Aeronautical Science*, 4:311, 1937; *Idem.*, 'The Spectrum of Turbulence,' *Proceedings of the Royal Society of London*, Ser.A. 164:476, 1938; A. A. Townsend, *The Structure of Turbulent Shear Flow*, London, Cambridge University Press, 1956; T. Von Karman and L. Howarth, 'On the Statistical Theory of Isotropic Turbulence,' *Proceedings of the Royal Society of London*, Ser. A, 164–192; 1938; G. N. Abramovich, *The Theory of Turbulent Jets*, Translation by Scripta Technica, Cambridge, Mass., M.I.T. Press, 1963.
37. B. R. Morton, G. I. Taylor and J. S. Turner, 'Turbulent Gravitational Convection from Maintained and Instantaneous Sources,' *Proceedings, the Royal Society of London*, Series A. No. 1196, Jan. 1956, pp. 1–23.
38. B. R. Morton. 'Forced Plumes,' *Journal of Fluid Mechanics*, 5, 1959, Part I,

pp. 151–163; *Idem,* 'The Ascent of Turbulent Forced Plumes in a Calm Atmosphere,' *International Journal of Air Pollution 1,* II, 1959, pp. 184–197; G. Abraham, 'Jets issuing into Fluid with Density Gradient,' *IAHR Journal of Hydraulics Division, Proc.,* Nov. 1961, HY 6, pp. 171–200; K. Cederwall, 'The Initial Mixing on Jet Disposal into a Recipient,' *Lab. Investigations,* Gothenburg 1963, Chalmers Univ. of Tech., Division of Hydraulics; L. N. Fan, 'Turbulent Buoyant Jets into Stratified or Flowing Ambient Fluids,' *Report No. KH–R–15,* June 1967, W. M. Kock Laboratory of Hydraulic and Water Resources, California Institute of Technology; W. Forstall and A. H. Shapiro, 'Momentum and Mass Transfer in Coaxial Gas Jets,' *Journal of Applied Mechanics* 17, 4, 1950, p. 399.

39. A. M. Rawn, F. R. Bowerman and N. H. Brooks, 'Diffusers for Disposal of Sowako in Sea Water,' *Journal of Sanitary Engineering, Div. Proceedings ASCE,* 86, SA 2, Paper No. 2424, March 1960.
40. G. N. Abramovich, *op. cit.* (see Reference 36).
41. W. Forstall and A. H. Shapiro, *op. cit.* (see Reference 38).
42. E. L. Bourodimos, *op. cit.* (see Reference 31).
43. U.S. Department of the Interior, Federal Water Pollution Control Administration, 'Industrial Waste Guide on Thermal Pollution,' Corvallis, Oregon, Pacific Northwest Water Lab., Sept. 1968.
44. *Ibid.*
45. P. B. Cheney, F. A. Smith, R. O. Brush, D. J. Pelton and J. D. Kangos, 'A Systems Analysis of Aquatic Thermal Pollution and Its Implications,' *The Travelers Research,* Corp. Report 7743 – 341 b, Prepared for the National Coal Policy Conference, Inc., Jan. 1969.
46. 'Industrial Waste Guide on Thermal Pollution,' *op. cit.* (see References 43 and 44).
47. American Chemical Society Report, 'Cleaning Our Environment – The Chemical Basis for Action,' *A Report by the Subcommittee on Environmental Improvement,* Washington, D. C., Committee on Chemistry and Public Affairs, 1969.
48. Guy-Harold Smith, *op. cit.* (see Reference 16).

Against Pollution:
Genetics

BJÖRN O. GILLBERG

Chemically Induced Genetic Damage

It has been known for about forty years that chemicals can induce genetic damage, i.e. mutations, in a wide variety of organisms, such as bacteria, fungi, plants, and mice. The genetic material being basically the same in all living organisms, such chemicals might, but do not have to be mutagenic in man. It is neverthless surprising, in these days of pollution, that it is not required by law in any country that chemicals directly used by man or spread out in his environment shall be proved to be safe from a genetic point of view, before their use is permitted. It is very important to make legislators all over the world aware of this problem, especially because modern man is exposed to thousands of synthetic compounds that did not exist ten years ago.

Mutagenic chemicals might not only hurt future generations but also contemporaries who get in direct contact with such chemicals, because, as is well known, mutagenic chemicals are often carcinogenic as well.

Man is today exposed to a broad variety of chemicals that are known to be mutagenic. For example sodium nitrite, used as a preservative (in bacon, ham, frankfurters, sausages, etc.), proves to be mutagenic in bacteria and other microorganisms.[1] Synthesis of nitrosamines from dietary nitrite and secondary amines in the stomach of animals and man has also been reported. Nitrosamines are carcinogenic in a wide range of animal species.[2] Optical brighteners that are used in cosmetics, detergents, soap, and even food in certain countries induce mutations in yeast.[3]

Certain pesticides, such as the widely used captan, induce mutations in neurospora molds and chromosomal aberrations in human tissue cultures. Air pollutants, such as benzopyrenes formed when burning coal, gasoline, etc., are mutagenic in mice. Nitrogen oxide from automobile exhaust can be converted to nitrous acid, which is mutagenic in many microorganisms.

A great number of drugs are also known to be mutagenic, among them streptonigrin and actinomycin D. Antidepressive drugs based on

213

hydrazinederivatives have also been shown to induce genetic damage in bacteria, broad-bean roots, and tissue cultures.[4]

Many chemicals used in the chemical industry are well-known mutagens. Take for example the epoxides and the di-epoxides that are used as basic material for the production of cosmetics, detergents, and lacquers or polyethyleneimines that are used in the fiber and rayon industry. It is very rare that these chemicals appear in the finished products, but their production and working with the products is connected with notable medical risks.[5]

The question of when to ban a suspected mutagen is controversial. Shall the society wait until a chemical has been proved to be mutagenic in a wide range of organisms? Or is there sufficient reason to ban a chemical that has been proved to be mutagenic in a few different organisms? In my personal opinion compounds that are mutagenic in microorganisms, fruit flies, or tissue cultures, should be banned and classified as potential mutagens in man, especially if there is no, or little benefit for the society connected with the use of the compounds. Man is today exposed to many chemicals when the risks of their use outweigh the benefits: Optical brighteners and food cosmetics (food colors and flavors), for example, used for the benefit of the seller rather than for any benefit of the consumer.

CITED REFERENCES

1. H. J. Sanders, *Chemical and Engineering News* 47, No. 23, 1969.
2. B. S. Alam, I. B. Saporoschetz, and S. S. Epstein, *Nature 232*, 1971, p. 116; W. Lijinsky and S. Epstein, *Nature 225*, 1970, p. 21; V. J. Sander, *Arzneimittel Forschung* 19, 1969, p. 1091; N. P. Sen, D. C. Smith and L. Schwinghammer, *Food Cosmet. Toxicol.* 7, 1969, p. 301.
3. B. O. Gillberg and J. Aman, *Mutation Research* (in press).
4. B. A. Kihlman, *Mutation Research* 1, 1964, p. 54; H. J. Sanders, *Chemical and Engineering News* 47, 1969, p. 50.
5. Ehrenberg and A. Gustafson, *A Translation of a Report of February 1959 to the Swedish National Board of Health*, 1970.

ØISTEIN STRØMNÆS

The Impact on Human Genetics

It can be safely assumed that there is general agreement regarding the basic processes involved in human evolution, and — more specifically — in the role played by the main forces in evolution: mutation and selection. The subject of genetics was born, or to be correct, reborn, in 1900, when it was found that mutations could occur, and that changed characteristics could be transmitted to future generations. The relative proportion of individuals exhibiting various characteristics in any given population will, of course, depend on the relative frequency of such characteristics in the parent population. The British mathematician Hardy and the German physician Weinberg discovered independently, in 1908, the mathematical expression which relates the proportions of characteristics existing in one generation to those in the next generation. A state of equilibrium will, of course, depend on the mutation rates on the one hand and on selective forces on the other. A particular defect may, for example, reduce the chances of mating between those individuals who exhibit this defect.

Any change in the mutation rate will alter the equilibrium among various characteristics in a given population. Once it is known that a human activity increases mutations, we can then either improve the chance for survival of individuals with defective inheritance and/or to correct the condition caused by it. In the more or less recent past, we have introduced an increasing number of mutagenic, i.e. mutation-causing, agents into our environment, such as ionizing radiation and a great many chemicals.

In 1927 H. J. Muller discovered the mutagenic effects of ionizing radiation, and he received the Nobel Prize for it. But it was not until twenty-five years later, in the fifties, in the aftermath of nuclear bombing and as a result of nuclear testing, that this subject received any public attention. Endless arguments ensued and continue to this very day. It was not enough for the mutagenic effects of ionizing radiation to be demonstrated in other life forms; scientists were expected to prove beyond a reasonable doubt mutagenic effects in human beings — and that was difficult. It took many years.

There were other difficulties. For years scientists in the US Atomic Energy Commission withheld secret data and insisted that fallout would be distributed evenly all over the world. Scientists had to go out and prove that fallout will concentrate more on the West Coast and less on the East Coast, or that certain radioactive elements will accumulate in specific plants or in specific organs, such as strontium-90 in milk and bones. I do not think that it would be unrealistic to say that the protests by an increasing number of scientists, who accumulated such data in many instances against great odds, at least contributed to the nuclear test ban treaty's finally having been adopted in the USA.

We now find ourselves in a similar situation regarding chemicals. That certain chemical compounds could be mutagenic was discovered in 1944. A good deal of time has elapsed since then and the potential danger to human beings from food additives, drugs, and other chemical substances has been demonstrated beyond question. The answer seems simple enough: if certain chemicals are toxic, or may cause cancer or mutations in man, then their consumption, or the exposure of human populations to them, should be prevented. However, if agreement is sought on the standards to be applied, difficulties arise immediately. If the concentration of a substance has been reduced below a certain level it may no longer be toxic to a man weighing 140 lbs, and yet it may still be mutagenic. At what level may a residual concentration of this substance in food or in the air be considered 'safe'?

Once we consider the effect of dangerous chemicals on the human population, it is not enough to be concerned only with as yet untested materials. We must investigate systematically possible somatic or genetic damage which may already have been caused by materials to which human populations have been exposed in the past, even when the materials in question have already been withdrawn from use. This applies for example to a variety of food dyes. Answers to such questions are most pressing because there appears to be a statistically significant increase in the frequency of a number of diseases, such as heart disease, lung cancer, emphysema, leukemia, and certain skin diseases over wide areas of the world.

It has been said that on the basis of currently available evidence the incidence of lung cancer and emphysema could be reduced by perhaps 90 per cent, if people were just to stop smoking cigarettes. At this point the questions arises as to what would be reasonable criteria to determine when the evidence is sufficiently compelling for health considerations to outweigh any possible economic arguments. On the other hand, the data presented by Sternglass at this conference regarding the possible effects of low-level radiation would demand drastic revisions of presently accepted international standards of radiation safety.

Given sufficient pressure it appears that national food and drug administrations demand more and more testing for toxicity and mutagenicity of foods, food additives, or drugs prior to their distribution and sale. But much uncertainty remains regarding appropriate standards of safety which ought to be established by law. There remains the questions of other chemicals (i.e. neither food nor drugs) which may have farreaching effects on human populations. To what extent should industry be responsible for conducting necessary tests before such products are put on the market?

The basic questions, then, which must be answered are: Whose responsibility should it be to demonstrate whether or not a chemical may be dangerous to humans? Should it be done by national governmental agencies? Or should it be required of the industry which produces and wishes to introduce a new product? Or should it be done by a system of random testing by independent scientists?

In conclusion let me say that I believe it to be the responsibility of the scientific community:

1. To press for the adoption of national legislation which would require testing of all new chemicals intended for human consumption to which human populations would be exposed, for toxicity or possible carcinogenic or mutagenic effects, prior to their being marketed.

2. To press for a systematic program of continuing research to determine the short- and long-range effects of chemicals to which human populations are now or have in the past been exposed.

AGNAR P. NYGAARD

The Ethical Problem of
Human Genetics

It is important to build up public pressure against pollution. I would like to draw your attention towards building up pressure against the misuse of science within the scientific community. In the last century, men of science firmly believed that science and technology automatically would increase the freedom of man. Now we are not so sure. We realize that sometimes dangers rather than blessings may be in store.

Dangers are becoming obvious in the fields of molecular biology and embryology where man has begun to tamper with the mystery of life itself.

We all want to be healthy and we all want to be gifted. It is the ultimate goal of eugenics to make *all* people healthy and gifted. But what will be the price we must pay for interfering with the natural course of reproduction?

More and more genetic defects can now be detected in the early embryo. We can detect, and possibly prevent the birth of mongoloid children and children with a number of other defects as the means of reducing human suffering and easing the burden of social expenditures. There is, however, another side to the problem. By letting the laboratory and the computer decide who is to be born, will we cause a cult of the gifted and a devaluation of the less able? Do the less gifted lack the capacities of 'selfrealization' and 'happiness'? And who is to decide who ought to be born? Who ought to live?

There is an increasing ability to fertilize eggs in culture, to grow them to the blastocyst stage and then implant them in a woman to grow to full term. Although this technique is not yet fully investigated or perfected in mammals, a method with human eggs has already been introduced by Robert Edwards at Cambridge. My guess is that it will take a lot of experimentation with human embryos at differing stages before the technique is perfected.

218

The intention, of course, is good. One will be able to by-pass a blockage in the Fallopian tubes and thus help some women to have the children they want. Brutalization may, however, be the result of experimentation with human embryos without due consideration of their human character.

We should be aware of the possibility that the technique introduced at Cambridge may later be combined with further manipulation of the embryo. I would like to mention here the possibility of adding to the embryo. There are already in existence reports to the effect that, in mice, donor cells can be injected into the embryo. The multiplication of the donor cells can lead to the partial colonization of many organs, including the brain. Thus, in future, we may be able to obtain 'hybrids' of man. In this case, there may also be health benefits in the sense of counteraction of genetic defects.

Where human beings are concerned, however, more than a question of health benefits is involved. How about the identity problem, the right to have a father and a mother, the right to be genetically correlated to our familial parents? Can we treat embryos indiscriminately as laboratory objects?

A scientific advance can either have beneficial or harmful consequences. We all agree that it is important not to stop the sciences of genetics, molecular biology, and embryology. If fundamental research in these fields is curtailed, we would slow the current progress in the recognition of genetic defects, the control of cancer, the improvement of livestock and crops. And last, but not least, we might prevent our progress toward the basic understanding of life.

In the interest of both science and humanity, therefore, we should lend thought to the proposal of James Watson, the co-discoverer of the DNA double helix, to form an international commission to control the course of human genetic experimentation. We ought to have a medical code of ethics which includes experimentation with the human embryo.

Against War

W. C. DAVIDON

Chemical and Biological Warfare: Pollution by Design

I want to speak briefly about two matters: pollution by design rather than as a byproduct of industrialization and what is to be done about it. Generally, after examining some of these problems, people say there is a need to build public pressure. I think this is inadequate, because most of us, in fact, are only beginning to learn about effective ways of changing policies. Talking generally about building public pressure does not, I think, tell us much about what is to be done.

Pollution by design

As far as pollution by design is concerned, many people have the impression that this is not, in reality, occurring. I think it is important to realize that we are not talking about a hypothetical situation or about plans for some future occurrence, but about existing policies of the major powers, particularly the policies of the US Government and the larger corporations which manufacture the materials used by it.

In 1968, the Society for Social Responsibility in Science sent a small team of scientists to Vietnam to make some preliminary analysis of the ecological effects of the large-scale use of chemical agents there (see p. 230). The study is incomplete, but it served to alert us to the problem and to spur the American Association for the Advancement of Science to conduct a more detailed investigation two years later.

The SSRS study clearly shows the magnitude of the uses of chemical agents being deliberately deployed to destroy the ecology of an area covering large parts of Vietnam, eastern Cambodia, and Laos. Millions of tons of highly toxic materials have been dropped over these areas.

I want to refer briefly to some statistics cited in the SSRS Report.[1] For example, between a half and a fifth of all the mangrove forests — covering several millions of acres in Vietnam — have been destroyed. And they have been destroyed permanently or, at least, for the foreseeable future. There is no indication that normal vegetation is returning to some of these areas. The sizable use of 2,4-D and of 2,4,5-T as agents to remove leaves continues, although the US Government has made statements concerning the restriction of this use. But according to research — independent of government pronouncements — into such sources as purchase orders that the government continue to issue for the acquisition of such devastating materials, and information obtained from American personnel in Vietnam who are opposed to US Government policy, one must infer the continued use of defoliants and similar agents. The only change consists of some of the actual spraying operations having been transferred to the Saigon Army. That is one method used by the US Army to state (technically) that the US Government is decreasing the use of defoliants, even when the actual use of these compounds is not *really* decreased.

In addition to the use of these toxic chemicals, there are sizable ecological effects from the massive bombardment that has taken place in Vietnam, Cambodia, and Laos. Even though this bombing does not consist of chemical and biological weapons, it must not be ignored. This is the heaviest aerial bombardment in the history of warfare — much greater than the heaviest aerial bombardments of World War II. For example, in 1968 alone there were over 2.5 million bomb craters produced in Vietnam. Many of these are of an average diameter of fifteen meters or more. These numerous bomb craters generally fill with water and become breeding grounds for insects, thus producing further irreversible changes in the ecology of Vietnam.

Some five million acres have been sprayed with defoliants. Arsenic compounds have been sprayed in large quantities in Vietnam: they have as their primary objectives the destruction of rice crops. The USA is trying to cut off the food supply for the population, particularly the population not under USA and Saigon control. In 1970, the so-called 'orange agent' was used to damage close to 200,000 acres in East Cambodia.

We can go on to cite further statistics, but I shall stop because my point is not to provide an exhausive indication of what has been the case — and still is — but to establish the fact that it has been deliberate policy of the US Government to change the general ecology of Indochina.

Some people have the impression that this policy has been revised. I shall, however, indicate what information exists to show that that is not the case. I shall quote from a report submitted by a group that is part of the American Friends' Service Committee.[2]

On July 13, 1971, the US Army announced the beginning of this projected, year-long destruction of stockpiled germ-warfare agents. The announcement followed a year and a half of periodic White House and Pentagon news releases repeating the theme set by President Nixon in November 1969, namely, that germ-warfare agents would be destroyed. A year and a half ago, Nixon also announced that the USA was renouncing the first use of lethal chemical weapons and all use of biological warfare agents. These Nixon announcements, coupled with the widely publicized and controversial dumping of obsolete nerve gas, have left the public with the impression that the entire US program of chemical and biological warfare is being halted.

While this impression is being given publicity, military contractors are privately being assured that chemical and biological weapons work is continuing, and even expanding, and that cutbacks involved only a tiny part of the overall program. A prominent defense market analyzing firm* has reported that work in the field of chemical and biological weapons research is continuing 'at funding levels equal to or exceeding those prior to the public relations announcements of cessation of these efforts.'[3] A limited-circulation intelligence report published by Defense Market Survey, a subsidiary of McGraw-Hill, states that 'despite public announcements to the contrary, the military agencies are not discontinuing chemical and biological warfare research.'[4]

The report continues to discuss the magnitude of the contracts issued, in what fields and where the contracts are to be sought. Thus the contamination of an environment is *designed*.

I think it is important to recognize that the disposal of chemical agents is a serious problem. In the USA, there was considerable attention paid to one shipment of some of these agents that were to be destroyed about a year ago. What was not as well publicized was the fact that this particular shipment, which was indeed stopped, was only one of a series of such shipments, and that the previous four or five shipments of similar magnitude had been transported to, and dumped into, the ocean without public comment. Emphasis on the success of stopping *one* transport has blurred people's understanding of the fact that this transportation was only one instance of a continuing policy. One shipment was stopped. How many have been and are still being dumped into the ocean?

Again, to get some idea of the magnitude of this disposal of poisons into the ocean — into international waters — let us examine the one particular shipment that did become public knowledge. It was probably

* These firms are engaged in the analysis of war expenditures for the purpose of making that information available to corporations seeking military contracts. Ordinarily these reports are kept confidential for the use of corporate leadership. But it is possible to get access to such reports. Much of the information about the continuing programs comes from study of these reports.

typical, consisting of a total of some 27,000 tons of material shipped in 809 railway cars. It contained 12,000 tons of GB, a nerve gas which the Nazis had called 'sarin'. It was only produced in small quantities by the Nazis, but it has been produced in very large quantities in the USA.

While some nerve gas was destroyed, so-called binary weapons have also been developed. A binary weapon is a substance stored not as one nerve gas, but as two separate liquids. Neither of the two liquids, by itself, is a nerve gas. But mixed together they become nerve gas. The two liquids are only mixed at the time of use as nerve gas against people. The two separate materials then, officially neither being a nerve gas alone, can be stored in large quantities. The US Government thus continues to produce materials for use as nerve gas, but can, with (technical) truthfulness, claim that the production and storage have been discontinued.

In addition to the 12,000 tons of GB, the one shipment (that has become public knowledge) also contained some 5000 tons of mustard gas. Asked why this material was being dumped into international waters, the Assistant Secretary of Defense of the USA, C. L. Poor, said that, had the material been disposed of in deep caves, we would have to be deeply concerned about the possibility of seepage into underground water supplies. This, he said, might have severe long-term environmental effects![5]

One example where the dumping of nerve gas had very visible, though not widely publicized, consequences occurred near Lake Island in the Pacific. The nerve gas that was dumped there some twenty-three years ago spread. About three years ago, the US Coast Guard station on the island had to be evacuated because of the increasing levels of gas in the area.

What is to be done?

My second subject is what one might do about the above problems. They can be stated at various public meetings, of course, but simply stating facts publicly — often at only small meetings of concerned people — does not change public policy. It is important for us to realize that exposure alone is not enough to influence policy.

Other methods need to be considered. Those of us who are teachers have a very real responsibility to change the curricula, not only for students of science but for students at all levels and in all fields. We must see to it that awareness of the scientific as well as the economic and political aspects of the situation be brought to the attention of our students. The notion of a narrowly defined technical or scientific or humanist education that deprives people of the opportunity of being participants in decision-making in their own communities is, I think, an outmoded concept. Education should mean not only talking about problems and decisions in general, but discussing them seriously when the curriculum is being established at one's own university. It is in the choosing of the subject matter for one's own course that one can actually produce change.

Helping to spread awareness also means reaching out and making more serious contact with people in industry. The members of scientific groups concerned with these problems tend to be predominantly from the academic world. A group of us, in the Philadelphia area, went through the membership lists of several scientific societies and wrote to those members who were affiliated with some of the large corporation in the area, suggesting that we meet with them regularly to discuss some issues of common concern. We got favorable responses from a very sizable group. To set up meetings between scientists in the academic community and in industry for the purpose of discussing topics ranging from job security to other questions of concern to the industrial scientist — which may differ from those that are of concern to scientists in the academic world — and to arrange these meetings on a continuing schedule, is the kind of communication between the academic and the industrial areas I think necessary.

It is important to realize that many scientists, particularly in industry, are very fearful of engaging in public action, because of their job insecurity. In this regard, Ralph Nader is keenly aware of the need for professional societies and groups to create a climate in which those engaged in industry can speak forcefully even on 'unpopular' subjects. Dr. Lenz, for example, has stated the difficulty of obtaining sales figures for drugs. If more scientists, technicians, and engineers in drug firms thought that they could speak about matters of public concern without endangering their livelihoods, such vital information might become more easily available. Many of us think that Daniel Ellsberg rendered an important public service in making information about the Pentagon war plans public. There are many potential Daniel Ellsbergs in many corporations, in many government positions, not only in the USA but in many countries. If we can build a climate in which such actions are looked upon as generally desirable — as actions that would have public approval and support — information that is of vital importance to the public might become publicly available.

Taking the initiative in getting such information is, of course, desirable, but there are also other initiatives that can be taken. One forceful action that the scientific community might take — and it is only rarely resorted to and even then on a very small scale — is the notion of a strike. Scientists might join together in a refusal to engage in their usual scientific activities, at least for a short period of time. This would both dramatize the urgency of our problems and allow scientists time for a systematic investigation of them. In the USA, on March 4, 1969, a sizable portion of the scientific community chose to discontinue the normal routines of their laboratories and teaching in order to focus attention on some of the public consequences of scientific activity. Choosing a day or a week, joining together with the other groups of people on a large scale, would serve to emphasize the most important issues.

There are, of course, other forms of direct action. Specific knowledge of the locality and circumstances of each problem is important and varies in each case. I shall cite, as an example, direct action at Dow Chemical Corporation, which has a large plant in Michigan. A great deal of research on chemical weapons, particularly on nerve gas, had been done at that plant. Results of the research were stored on magnetic tape in the computing facility at the plant. A group of people went to that facility (at a time when the plant was not in use) and with strong permanent magnets erased the information stored on the magnetic tapes. (A permanent magnet passed back and forth close to magnetic computer tape will erase most of the information stored on the tape.) Such extreme action against research that seriously endangers human life is called for in the crisis of our time.

To summarize, some of the possible kinds of actions are:

1. Organizing courses at universities and action at universities to introduce subjects of public concern to a broad spectrum of people.

2. Reaching out to scientists, technicians, and engineers in industry in one's own area for sustained discussions of subjects of public and specific concern.

3. Trying to protect individuals in industry or elsewhere who speak out on matters of public concern and who take the initiative in getting access to otherwise unavailable information pertinent to public health and safety.

4. Using a strike or temporary work stoppage as a way of dramatizing issues of public concern as well as for the purpose of rallying people or providing time for competent people to study such problems.

5. Taking direct, even extreme action that builds concern for human life, when such action appears to be the only solution.

CITED REFERENCES

1. SSRS Mission to Vietnam, *Scientific Research* 4, June 23, 1969.
2. Art Kanegis, *US CBW Policy.* NARMIC Report, American Friends Service Committee, Philadelphia, Pa.
3. *Ibid.*
4. *Ibid.*
4. Marc Lappe, 'Chemical and Biological Warfare,' Student Research Facility, Berkeley, 1969, p. 19.

AN INTERVIEW

SSRS Mission to Vietnam

For years members of the Society for Social Responsibility in Science
(and others) had urged that an American or international team of scientists brave the dangers of war-torn Vietnam to study the effects of defoliants on the ecology. But because of the apparent impracticality of
doing research in a battle zone, and the refusal of both the US Department of Defense and the United Nations to endorse such a mission to
Vietnam, nothing happened.

When the Society for Social Responsibility in Science decided to send
two scientists to Vietnam, Egbert W. Pfeiffer of the University of Montana and Gordon H. Orians of the University of Washington were chosen for this first and modestly financed (by individual members of the
SSRS) mission, designed as a pioneer effort to stimulate further and more
extensive work later.

Orians specializies in the study of the evolution of vertebrate social
systems, especially the ecological factors that determine the number of
animal species an environment will support. Pfeiffer is a vertebrate
zoologist interested mainly in the renal physiology of mammals.

They flew several (U.S.) defoliation missions and inspected the effects
of defoliants from helicopters, from armoured gunboats in the Rung Sat
area of the Saigon River, and on the ground in company with Vietnamese biologists and French rubber planters.

In this interview they talk mainly about their experiences in (and
over) the countryside.

Why did you go to Vietnam and how much were you able to see and do?

We visited the Republic of Vietnam on behalf of the Society for Social
Responsibility in Science from March 17 to April 1, 1969, for the purpose of gathering information about the ecological effects of the war.
Particularly, we wanted to study the consequences of widespread applications of herbicides to the forests and farmland of that country. We
gathered information through:

230

—interviews with U.S. military personnel,

—direct observation from a U.S. Army helicopter traveling over areas damaged by B-52 bombing raids,

—direct observation from U.S. Air Force C-123 aircraft, modified for spray application, accompanying U.S. military personnel on 'spray missions,'

—direct observation from U.S. Navy patrol vessels during a 2-hour-and-40-minute trip through the Rung Sat Peninsula on the lower Saigon River. The Rung Sat is an extensive region of mangroves that has been heavily defoliated,

—unclassified information, willingly supplied by U.S. military authorities,

—data in the files and publications of the Rubber Research Institute of Vietnam,

—a visit to the research station of the Institute, where we observed trees recently damaged by defoliants,

—pictures of trees damaged and killed by previous defoliations,

—data from the files of the Michelin plantation regarding the nature and extent of herbicide damage to rubber trees on Michelin plantations,

—interviews with the members of the faculty of science, at the University of Saigon,

—interviews with scientists in Vietnamese government agencies, including the Ministries of Fisheries, Forestry and Agriculture, which are staffed with biologists trained primarily in France and the United States,

—interviews and discussions with other Vietnamese.

Did you focus on specific areas of the Vietnamese ecology?

Since previous work in the field on the effects of defoliation has dealt almost entirely with direct effects upon plants, we made a special effort to observe animals in the area we visited, and to inquire as much as possible about changes in the animal population. Since our knowledge about birds is extensive, we learned most about that group of animals by our own observation; we gathered information about other taxonomic groups through interviews.

Although our visit was too short to obtain definitive answers to some of the most important questions that have been raised by the American scientific community about the ecological effects of the war, we feel that the material we gathered forms a significant contribution to continuing efforts to assess the impact of modern warfare on the environment in which man must live.

In your opinion, what are American scientists who worry about Vietnam most concerned about?

It is the widespread use of herbicides in Vietnam that has been of greatest concern to American scientists. We therefore gave top priority to learning about the effects of the defoliation program in Vietnam.

Defoliants have been used in Vietnam by the United tates since 1962. The program started modestly but increased sharply after 1965. A peak was reached in 1967; a slight reduction of areas sprayed with defoliants followed in 1968 as a result of the reassignment of equipment for other missions in response to the Tet offensive.

Where is the military conducting its spraying?

The bulk of the spraying is directed against forest and brush, but a significant proportion is directed against cropland in the mountainous parts of the country. U.S. military authorities believe the food grown in the mountainous areas is used to feed the forces of the National Liberation Front. They deny using defoliants on rice crops in the Delta region.

Much of the defoliation is along roads and rivers and around military establishments; border areas (near Laos and Cambodia) are also extensively defoliated. Forested regions north and northwest of Saigon have been very hard hit. This area contains some of the most valuable timberland in the country. In most cases, broad forest areas have not been repeatedly defoliated, though possibly 20 to 25 percent of the forest areas have been hit more than once. Roadsides and river banks are sprayed at regular intervals.

Who controls defoliation missions?

It is the official policy of the U.S. Army that all defoliation missions are initiated by the Army of the Republic of Vietnam, The final authorization for a defoliation mission is given by the U.S. Ambassador to Vietnam. The number of requests for defoliation missions greatly exceeds the present capacity of the Air Force unit that carries out the spraying missions, under the code name 'Operation Ranch Hand.' Therefore, priorities must be established in accordance with military considerations.

Whenever possible, aerial reconnaissance of the area is made before a final decision to spray. Evidently this is not always feasible, and some flights, including one we accompanied as observers, are flown over areas that were not checked out by the officers in charge.

232

When are the missions flown?

To reduce transport of herbicides by the wind and to improve the kill on the desired target, the military authorities have established regulations spelling out the conditions under which defoliation may take place. Missions are to be flown only when the temperature is less than 85° F and the wind is less than 10 knots. This restricts aerial spraying to morning hours, though usually an attempt is made to fly two missions each morning.

What defoliants are used and where?

Around Saigon, where drifting herbicides pose threats to cropland, the White agent is now preferred because of its lower volatility and greater persistence, but in regions where there is little agriculture Orange is the preferred agent because it is more economical. Orange constitutes about 50 percent of the total herbicide used, White 35 percent, and Blue 15 percent. Blue is used primarily against mountain rice crops.

Were you able to get out into the countryside?

It was impossible for us to visit defoliated forests on foot or by means of ground transportation. Therefore, we are unable to add much to what has already been reported on the direct effects of defoliants on forest trees. We can confirm the 1968 report of Fred Tschirley of the U.S. Department of Agriculture that the trees collectively known as *mangroves* are extremely susceptible to the action of defoliants and one application at the normal level employed in Vietnam is sufficient to kill most of the trees.

Most of the areas we visited by boat on the Rung Sat Peninsula were still completely barren, even though some of the area had been sprayed several years earlier. Only in occasional places was there any regeneration of mangrove trees. We observed no growth of the saltwater fern that often invades mangrove areas. The lack of fern in the area may be due in part to the extensive defoliation, which has removed potential seed sources.

What's happening to animal life?

As might be expected, the almost complete killing by herbicides of all vegetation in the mangrove areas has had a severe effect on the animals living there. During our tour of the defoliated areas we did not see a single species of insectivorous bird with the exception of barn swallows, which are migrants from the north.

233

This is what the military is spraying

Agent	Composition		Lb/gal	AE*	Use
Orange	N-Butyl ester 2,4-D	50%	4.2		*General defoliation*
	N-Butyl ester 2,4,5-T	50%	3.7		*of forest, brush*
		Total	7.9		*and crops*
Purple	N-Butyl ester 2,4-D	50%	4.2		*General defoliation*
	N-Butyl ester 2,4,5-T	30%	2.2		*agent—interchange-*
	Isobutyl ester 2,4,5-T	20%	**1.5**		*able with orange*
		Total	7.9		
White	Tri-isopropanolamine salt 2,4-D	2.0			*Forest defoliation—*
	Tri-isopropanolamine		0.54		*long term*
	salt picloram				
		Total	2.54		
Blue	Sodium cacodylate	27.7 %			*Rapid short-term*
	Free cacodylic acid	4.8 %			*defoliation—*
	Water, sodium chloride balance	3.1			*grass and rice*

* Active Equivalent of the acid.

Fish-eating birds seem to have suffered less severely, but even their numbers were much smaller than we expected. The birds we observed during a 2-hour period in the defoliated areas were: two oriental darters, 13 grey herons, three large egrets, 12 little egrets, one intermediate egret, six Javan pond herons, two storks, one black-winged kite, nine ospreys, three whimbrels, 10 little terns, and two white-breasted kingfishers.

All except the kite, which feeds on small mammals, are fish-eating birds. This suggests that aquatic food chains in the mangroves may have been *less* severely affected by defoliation than the terrestrial ones. The only other vertebrate we saw in the defoliated areas was a large crocodile.

Of all the areas in Vietnam, the mangroves in the delta of the Saigon River have probably been most severely affected by defoliation. The area treated is extensive, covering many square miles. The vegetation is extremely sensitive to herbicides, and since many of the species of animals inhabiting mangroves are restricted to that type of vegetation, they are more susceptible to extermination than species characteristic of the upland areas. Long-term studies of the Rung Sat ecology, including

234

the investigation of the status of such invertebrates as crustaceans and mollusks, should be given a high priority.

Were you able to get a good look at the upland forests?

Unfortunately, we were limited to aerial reconnaissance. So we have nothing to add to the published studies about the short-term effect of defoliants on tropical forest trees after single applications of herbicides. These studies deserve wider attention than usually given to the technical bulletins in which they have been published.

The area we observed from the air had been sprayed previously, and many of the trees in the target area of the mission appeared to be dead. Except for the wetter spots covered with bamboo, the ground in most areas was clearly visible from the low flying aircraft. Many areas in war zones C and D have been sprayed more than once.

Vegetative recovery was limited to the growth of bamboo and under-story trees, rather than to refoliation of the canopy diminants. The pilots could not tell which area had been defoliated more than once, so we were unable to compare the appearance of these forests to areas sprayed only once.

Limited observations of defoliated upland forest were made on the ground by Tschirley and by Barry Flamm, chief of the forestry branch of the Saigon office of the Agency for International Development. During his studies, Flamm visited defoliated forests near Special Forces camps in Tay Ninh and Binh Long, provinces northwest of Saigon, a region of grey podzolic soils.

According to these studies, there appears to be a modest kill of canopy trees on sites sprayed once, but understory seedlings and saplings survive, and forest regeneration begins fairly rapidly. However, on sites that received two sprayings separated by an interval of roughly one year, a heavy kill of all woody plants, including seedlings, is reported.

Two or three sprayings may kill approximately 50 per cent of commercially valuable timber in such forests. These areas are being invaded by grasses resistant to forest defoliants and they may prevent the reestablishment of tree seedlings for a long time; even if that does not occur, it will take many decades before a mature forest composition and forest physiognomy, may persist for even longer.

One year after spraying, timber is still in good condition, and could be harvested for commercial use if equipment and markets were available. However, shrapnel lodged in the trees will be a serious problem for the Vietnamese sawmill industry for many years. Many sawmills report that they lose from one to three hours each day because shrapnel in the logs damages saw blades.

What is the intensity of the sprayings, and are the applications uniform?

We had an opportunity to study the effects of unusually high rates of application of herbicides. For example, before jet pods were installed in the C-123 aircraft the planes were unable to remain aloft when engine trouble developed. In such a contingency, the crew was permitted to jettison the entire contents of the spray tank (1000 gallons) in slightly less than 30 seconds, whereas normal spray time is about 4 minutes. Although such contingencies are said to occur less frequently now, they do happen. On the spray mission that I (Pfeiffer) accompanied as observer, the spray nozzles of one plane failed to work properly, and the entire tank was unloaded at the end of the target.

Do we know the sites of such intense applications?

Since the locations of targets are pinpointed very precisely, and since reports are made of all unusual activities during a spraying mission, it should be possible to keep a record of such occurrences. And it is most important that all such incidents be recorded so that biologists will someday be able to investigate the sites of concentrated defoliant applications.

Have the Vietnamese been able to carry out any studies of their own?

Physiological studies of the effects of the defoliation of rubber trees have been initiated by the Rubber Research Institute of Vietnam, concerned because of the economic importance of rubber trees to Vietnam and because of widespread damage done to plantations by military spraying. Although these studies contain the best available data, they have been limited by the shortage of funds and the difficulties of field work in a country at war.

Is damage to rubber trees serious?

Damage to rubber trees in Vietnam has been extensive. The total yield of rubber and the yield of rubber per acre are declining. In 1960, rubber plantations in Vietnam yielded 2,345 pounds of dry rubber per acre (on plantations of more than about 60 acres or 25 hectares). In 1967, the yield dropped to 1,745 pounds per acre. In contrast, in Malaysia the yield in 1960 was 1,668 pounds of dry rubber per acre, but had risen to 2,215 pounds per acre in 1966. The decrease in yield is due to a combination of such circumstances as the cessation of tapping forced by military action, less experienced labor and less thorough control in the field, herbicide damage, lack of general upkeep of plantations, and the cutting of rubber trees along roads. The relative importance of each factor seems impossible to assess, but they are *all* consequences of the war.

The greatest loss over a one-year period has been estimated to be as much as 30 per cent of the normal yield. At current prices that amount of loss reduces profit from about \$220 per acre per year to zero. As a consequence, most of the smaller plantations have been unable to stay in business; only the large planters who have solid financial backing can weather such losses and remain in operation.

According to studies made by the Dow Chemical Co. (as reported to us by the Rubber Institute), the defoliant is absorbed through the leaves of the trees and it is carried down through the phloem within 24 hours. In fact, symptoms of defoliation may appear weeks after spraying. The distance the defoliant travels down a tree is a function of the dosage received, and the Institute people assess it by cutting into the trunks of the trees at different heights to investigate the flow of latex. Necroses are also clearly visible in the sectioned trunks, many of which we examined in the Institute laboratories.

As might be expected, the smaller the rubber tree, the more readily it is killed by defoliants. Research in Malaysia has shown that all concentrations of *n*-butyl ester of 2,4,5-T killed rubber seedlings in six weeks. Accidental defoliations in Vietnam indicate that trees less than seven years old can be killed by the dosages used in military operations, but older trees normally recover. Nevertheless, all the trees on 250 acres of Plantation Ben Cui were killed by herbicides in 1965, despite the fact that they were 33 years old. From such occurrences, the Rubber Research Institute concludes that repeated defoliations are creating a threat to the very existence of rubber culture in Vietnam.

In spite of this evidence, spokesmen for Chemical Operations, U.S. Army claims that rubber production is stimulated by defoliation. From our observations, although we do not claim expertise in this field, damage to rubber production is severe. The Rubber Research Institute, which only maintains experimental plantations and therefore cannot be accused of bias, is in an excellent position to conduct further research into the physiological effects of defoliants on rubber trees but needs funds.

What about other types of trees?

We were able to observe defoliation damage to other species of trees as well. On March 25, in the village of Ho-Nai we observed many fruit trees that had recently been damaged by defoliants. The characteristic sign was the presence of curled, dead leaves on the trees. Damage on the south side of the trees seemed excessive, suggesting that the spray was carried into the village by a southeasterly wind. Villagers informed us that spray had hit them about one week previously.

Chemical Operations, U.S. Army, reported to us that a defoliation aircraft had had to jettison its chemicals when taking off from nearby Bien Hoa Air Base at approximately the time when the Ho-Nai residents had observed the spray. The most severe damage was to jack fruit trees, which also produce a milky sap. The residents claimed to have been affected by defoliation missions no less than seven times within the past year.

How serious is the damage to these trees?

On March 23, in a residential area between Saigon and the U.S. Air Base at Bien Hoa, we examined and photographed many diseased mango trees. The owner of the trees, a U.S.-trained biologist, claimed that they suffered defoliation three years before, had become infected, and had not flowered or produced fruit since. In other areas we observed the same symptoms in mangoes and other trees. According to the Rubber Research Institute, latex-producing trees seem to be more susceptible than other species to herbicide damage.

Has anyone figured out the total damage to date?

Every Vietnamese biologist we talked to explained that herbicide damage has been frequent and regular over much of the delta region. In the Department of Agriculture we were shown photographs of damaged jack fruit, manioc, and rubber trees, and were told that many guava trees had been killed. The Department has attempted in a preliminary way to assess the total damage reported but finds it so extensive that adequate financial compensation to the owners of damaged trees would probably be impossible. The experimental station of the College of Agriculture of the University of Saigon at Tu Duc has been defoliated several times, most recently within the past month, usually with almost complete kill of vegetables.

What is the threat of herbicides to animals?

Previous studies have suggested that at the prevailing concentrations, herbicides are not *directly* toxic to animals — and we uncovered little evidence to the contrary. The Tan-Son Nhut Air Field in Saigon, for example, is sprayed by hand with agent Blue several times each year, but it has a serious rat problem, nevertheless. Every night a trapping crew puts out 100 snap traps and 30 live traps baited with bacon. From January 3, 1969 to March 19, 1969 they trapped 613 rats and eight small carnivores of at least two species. On two mornings we netted and observed birds in a brushy area near Bien Hoa and found them to be very numerous. We noted much territorial defense and singing, as would be expected at the end of the dry season in the tropics.

We received one report of many sick and dying birds and mammals in forests that had been defoliated, and two reports of the death of large numbers of small pigs near Saigon, but were unable to follow up the reports. The Ministry of Agriculture has received no bona fide claims of animal damage from defoliants.

Nevertheless, it must not be forgotten that habitat destruction, which defoliation regularly accomplishes, in most cases means death for animals. The widespread view that animals can move to other nearby areas is untenable; it runs counter to recent ecological evidence that tropical forests already hold the maximum populations of most species that the resources will support. Reduction of forest habitats will decrease the populations of forest animals by an equivalent amount.

Nor is it true that forest species can live successfully in the greatly modified forest conditions that prevail in even partially defoliated forests. They may have to wait until the basic food resources, such as insects and fruit, have built up again, and we do not know how long that will take.

How about fish?

There is a phenomenon that should be investigated immediately: it is a widespread sickness that appears at the beginning of the rainy season in commercially important fresh-water fishes. The symptoms are many small, round dark spots in the muscles and an adversely affected taste. Poor people, however, continue to eat the diseased fish. This disease has characteristically appeared at that time of the year in Vietnam, but the director of the Institute of Fisheries has received reports suggesting that its incidence is now higher than ever. It should not be too difficult to determine the cause of the disease and find out whether it is, in fact, more widespread in areas subjected to more intensive defoliation than in areas only lightly affected.

What is happening to the soil?

Because neither of us is trained in soil science, we lack the competence to assess the problem of rock decay as a result of defoliation. Soil scientists in Vietnam have not received any reports of soil toxicity nor of any decline in fertility that could be attributed to herbicides. The areas most likely to be affected are not presently under cultivation, however, and since they are not controlled either by the South Vietnamese or the American forces, scientists from Saigon government agencies or the University of Saigon cannot investigate. Since rock decay should be a long-lasting effect, there would presumably be no difficulty in determining its extent at a future date, but a delay may prevent the assessment of shorter term persistence problems.

Direct and deliberate application of agent Blue to cropland has been restricted to the highland regions of the country, which are held by the North Vietnamese. Consequently, Saigon and American scientists are unable to make studies there, either. American officials consider the agent Blue program very successful because many captured NLF soldiers from those areas are seriously undernourished. Some are even stretcher cases at the time of capture. This so-called 'resource-denial' program makes the efficacy of herbicides evident.

Is the heavy bombing doing any long-term damage to the ecology?

Though it has not attracted the concern of American scientists, the damage caused by B-52 bomber raids is of considerable ecological significance. The 500- and 750-pound bombs dropped by these aircraft leave craters as much as 30 ft deep and 45 ft in diameter. Most of them are filled with water, even late in the dry season.

The U.S. Army does not disclose the total number of bombs dropped, and the total area affected cannot be calculated accurately. However, the magnitude of the effect can be estimated from the following facts: the standard load for a B-52 is 108 500-pound bombs, or nearly 30 tons of explosives; and a 'mission' normally consists of 3 to 12 aircraft. In 1967. 982 missions were flown over the Republic of Vietnam. In 1968, 3022 missions were flown. The number of missions now being flown is greater than ever before. Assuming an average of eight planes per mission, one can estimate that about 848,000 craters were formed in 1967 and 2,600,000 craters in 1968. As one Vietnamese put it, we are making the country look like the surface of the moon. Unless heavy earth-moving equipment can be brought to the sites to fill the craters, they will remain a permanent feature of the South Vietnamese landscape.

Obviously these craters are potential breeding grounds for mosquitoes. They may possibly be made into fish breeding ponds. They may also render many agricultural areas difficult to utilize. The potential significance of the craters is so great that intensive studies and appropriate follow-up action are urgently needed.

240

What else struck you about the situation there?

The prolonged military activity in Vietnam is also causing other up-heavals, including major sociological changes. One dramatic example is the amazingly rapid rate of urbanization as people flee the war-torn countryside or are forcibly transported to the city. Within the least decade, Saigon has changed from a quiet city of 250,000 inhabitants to an overcrowded city of 3,000,000. This and the tremendous infusion of American capital have caused a rapid increase in the number of motor bikes and small cars in the streets — and Saigon's traffic accidents have become notorious.

The air pollution problem in the city is so severe that many trees along Saigon's major arterials are dead or dying. It is possible that the wind drift of defoliants has contributed to weakening the trees, but it is likely that the major cause of this injury is air pollution.

Is anything being done to solve these problems?

There are no immediate prospects for improvement. Creation of an adequate municipal transportation system to ease the traffic of private vehicles seems as improbable in Saigon as it is in most American cities.

And outside Saigon, did anything else strike you?

A major cause of forest destruction in Vietnam today is fire. Some fires are started deliberately by Vietnamese forces and some are caused by artillery shells. Over 40 per cent of the pine plantations in the country has been burned out recently; how much of the mixed forests has been destroyed by fire is unknown.

Because of the war, all hunting in the Republic of Vietnam has been officially discontinued. The beneficiaries of this ruling are tigers. In the past 24 years, these animals have learned to associate the sounds of gunfire with the presence of dead and wounded human beings. As a result, tigers are drawn by the sound of gunfire, and apparently consume many of the human battle casualties. Although there are no accurate statistics on tiger populations of the past or present, it is likely that tigers are multiplying in South Vietnam, much as the wolf population increased in Poland during World War II.

What conclusions have you reached about the effect of chemical warfare?

In Vietnam, the chemical weapons of a technologically advanced society are being used massively for the first time in a guerrilla war. In this war there are no battle lines, no secure territory, and no permanent military installations that can serve as targets for attack. Rather, the military efforts are aimed at increasing the toll of fatalities, denying food to the enemy, and depriving him of the cover and concealment afforded by natural growth.

The effectiveness of using defoliants is evident. Our own observations showed the profound results of denuding the country of growth. The military is emphatic about the effectiveness of defoliation. They claim it is significantly reducing American casualties. Therefore, the demand for the services of 'Ranch Hand,' the code name for the defoliation program, greatly exceeds its ability to supply them.

Although the total number of requests for defoliation missions was not disclosed, we were told that even if no further requests were made, the defoliation crews would be kept busy for years by the present backlog. The current level of the defoliation program is not determined by military demand nor by any consideration of the ecology and viability of the land and the conservation of the natural resources of Vietnam, but solely by competition for equipment and personnel.

Do you consider the consequences to be severe?

We consider the ecological consequences of defoliation to be *very* severe. Enough is now known to reveal that a significant fraction of mature trees in most forests is killed by a single application of herbicide and that almost complete kill, including destruction of seedlings and saplings, is to be expected if repeated sprayings are made. Because of military demands for respraying, we must expect the virtual elimination of woody vegetation at defoliation sites as a common result of the military use of herbicides.

Acres sprayed per year*

Year	Defoliation	Crop Destruction	Total
1962	17,119	717	17,836
1963	34,517	297	34,814
1964	53,873	10,136	64,009
1965	94,726	49,637	144,363
1966	775,894	112,678	888,572
1967	1,486,446	221,312	1,707,758
1968	1,297,244	87,064	1,384,308

* These are estimates based on number of spray missions flown, calibrated rates, and width of spray swath.

But is this not limited to certain areas?

It is probably so intended. But it is evident that the most stringent regulations for the application of defoliants to a certain area cannot prevent the widespread dispersal of herbicides to areas far beyond those that were intended for defoliation. We found abundant evidence of repeated moderate-to-severe defoliation in areas many miles away from the sites of direct application. Every responsible Vietnamese we met confirmed this.

A pilot in a war zone will jettison his load of defoliant rather than jeopardize the safety of his crew and plane; and a spray plane will not return to its base with a full tank just because its crew found the temperature or the wind velocity higher in the target area than anticipated. The military use of defoliants will inevitably result in herbicide damage to areas that are far more extensive than the specified targets.

How do the Vietnamese people react to the defoliation program?

It is evident that thedefoliation program has had tremendous psychological impact on the Vietnamese people and that it has profoundly affected their attitude toward Americans. A farmer whose entire crop has been destroyed by herbicides and whose fruit trees do not bear for three years is bound to be full of strong resentment. We were told repeatedly, though politely, that attitudes toward Americans have deteriorated significantly as a result of the massive use of defoliants.

The U.S. claim that defoliation is more humane than other weapons of war because it does not cause human casualties directly may appeal to those whose land has not been defoliated, but hardly to those whose food supply or property has been destroyed.

A realistic assessment of the effects of defoliation must take into account the psychological effects on the population, which — in this kind of war — the defoliators claim to be 'protecting, liberating, and pacifying.'

How do you feel about the military's role in all this?

The politically sensitive nature of defoliation is fully recognized by the military authorities. Although they claim that defoliation produces no long-term effects on the ecology, they have instituted the most stringent regulations to govern herbicide use. The Army claims that it is more difficult to get permission for the defoliation of trees in Vietnam than for killing people, and permission to spray rubber trees directly has never been granted, according to military sources, even when enemy forces were 'known' to use plantations for concealment. Preferential

243

treatment of the politically powerful rubber interests in Vietnam seems to have added to the hostility of the poorer Vietnamese.

The secrecy surrounding the use of defoliants in Vietnam has also contributed to the feelings we have reported here. Neither the government of the Republic of Vietnam nor American officials have disclosed information to the Vietnamese about the agents used, the areas sprayed, or the nature of the chemical action of defoliants and herbicides. The most concerned Vietnamese scientists did not know the chemical compositions of the herbicides, even though they have tried to get this information from their government.

Where do we go now?

American scientists will want to know what investigations might be immediately possible to sift the facts from among so many conflicting claims regarding the ecological effects of defoliants, and to stem the tide of increasing mistrust between the Vietnamese and the Americans. Support for research projects should be initiated by the American scientific community without delay. There are scientists in Vietnam who have been well trained at American and European universities and who are deeply concerned about the effects of the war on the ecology of their country. They are eager to conduct the research programs that are necessary for the rehabilitation of their ravaged land.

The flora and fauna of the country are well known. The Rubber Research Institute of Vietnam continues to function even though military action has forced it to relocate. It is capable of conducting research on the physiological effects of defoliants on rubber trees and other species. The Institute's staff is interested in investigating the possibilities of diversifying so that it can advise rubber planters on how to avoid their complete dependence on rubber.

Vietnamese biologists are well qualified to do significant research. A modest investment of funds is likely to produce important research results and would also improve Vietnamese relations with American scientists.

Although long-term studies, such as following vegetational succession on heavily defoliated areas, would be impossible for Saigon or American investigators to carry out, there are no insuperable barriers to the investigation of fish diseases, of methods of minimizing herbicide damage to commercially important trees that have been deliberately or inadvertently sprayed, or of further animal toxicity studies. It should also be possible to gather soil samples from areas that have been subjected to different treatments to learn more about the fate of arsenic compounds and the possible accumulation in the soil of the more persistent herbicides.

We urge that such studies be initiated now, rather than be delayed until hostilities cease, even though the difficulties are obviously great. We recommend most strongly that the American Association for the Advancement of Science, in accordance with its resolutions of 1966 and 1968, take the initiative in setting up an international research program on the long-range effects of the military use of herbicides in Vietnam.

We believe that such action is necessary if U.S. scientists wish to maintain — or regain — the respect of scientists in Southeast Asia.

Toward Education

VICTOR PASCHKIS

Education for our Changing Technology

If we want to control and limit pollution we need immediate changes in education on two wholly different fronts. First must come sophisticated training in the life and physical sciences to determine for every technical process the amount and nature of pollution, as well as technical means to decrease it. The urgent need for such education is widely acknowledged, yet often neglected.

The second need is less understood but equally urgent. It arises from two facts: first, that every item man produces adds to the pollution of the world — pollution from any technical act can be decreased but never eliminated — second, that the world population is still increasing and we are far away from numerical stability. Thus it becomes necessary to assess all new products — and hopefully also all existing products — for *unintended* effects, with pollution high on the lists.

It is clear that vast educational changes are imperative, not only for scientists and engineers, but also for the community at large, on many levels. I will draw my examples from the USA.

Education to fulfill the first need cited above requires that for each process all effluents — solid, liquid, or gaseous — must be described quantitatively. Where applicable, the dependence of nature or amount of the effluents on weather or local conditions must be included. Since it is not to be expected that the engineer would understand the impact of these several effluents on the somatic or genetic conditions of the population, the instruction would have to show that it is necessary to seek the cooperation of physicians and geneticists.

Conversely, in the life sciences, the student must be led to study and understand the impact on health of unusual materials produced by technology intentionally or otherwise. Some advocate a special professional field of pollution control. Personally, I deplore this kind of specialization. It appears to free the design engineer of concern for the pollution which his design may cause — in use or during production. At

249

present design engineers are not trained to include consideration of pollution control in their work. It may be desirable as an intermediate measure to train pollution specialists until all practicing engineers will include pollution considerations in their work. Such specialists are trained now either in the departments of chemical engineering or in separate departments; such special study courses have recently been introduced in US schools under such names as Bio-Engineering.

Here we must introduce the new concept that an engineering task comprises more than achieving the stated purpose of the work. Since every technical task yields unintended effects, in addition to the intended ones, we must introduce into engineering education the concept that no engineering task is complete without study of the unintended effects.

Awareness of this is spreading rapidly in the USA; since it is a quite new concept the responses of different schools and teachers vary.

Some schools approach the problem from a historical angle and introduce courses in history of engineering, which then lead to the needed changes. From what I have seen of this approach, it seems an unfortunate use of time. Too much time is spent on the past, and the course is concluded by a brief consideration of necessary changes. However, I do agree that some reference to the past is helpful: but history should be introduced only as background for today's problems.

Other schools have introduced courses on Technology and Society. There are probably as many approaches as there are courses. In some instances, case studies are presented comparing solutions of engineering problems with and without inclusion of unintended effects. Others involve the students by assigning an individual or a group of students a design task. Under the leadership of the faculty member, the students are moved toward the inclusion of at least some unintended effects.

Still other courses discuss the nature of the unintended effects in a classification procedure (environment and pollution, using up of finite resources; social effects such as invasion of privacy, decision-making by computers). Besides the content of such courses, we have to consider both the teachers available and the level on which such courses should be taught.

I fear that a large number of teachers in US colleges reject the expansion of engineering in the direction indicated above. Frequently these teachers are those who also oppose the teaching of specific designs or of economic considerations of engineering design; they tend to view engineering problems simply as questions of applied physics and chemistry. Frequently, younger teachers and students urge introduction of such new courses, which in engineering education poses a difficult problem. While we cannot objectively blame the individuals for not having introduced this kind of reasoning in engineering work previously, it it almost inevitable from a sociological viewpoint to blame technology for many of the ills of today's society. My own answer to this rejection of technology is that it should be obvious to any scientist or engineer that mankind could not survive without technology and that we have reached the historical moment where control of technology is necessary. But the inevitable tone of some criticism in presenting new views is closely tied in with the question of *when* these ideas should be introduced in engineering education. In US universities and colleges, including engineering schools, the first one or two years are given to a largely general education plus science background. Specific engineering courses are concentrated mainly in the third and fourth years, or — if there is a fifth year — then.

You may hear arguments for putting a course in Technology and Society in almost any of the several years of an engineering school. People who want to place it in the early years claim that in the later years engineering students are spoiled by the narrow and outmoded technical approach taught by those who reject the broadened concept of technological responsibility. This stance is countered by teachers who believe that a meaningful discussion of pollution and other unintended effects of a process becomes valuable for students only in later years, when they have already had some technical training.

Ideally, one could argue that there should be one course every year, however short, dealing with this huge problem. Some will say that engineering has become so complicated that four or five years of engineering school are insufficient, and that we dare not take time from 'pure' engineering subjects. I believe that this argument is untenable: increasing technology makes it possible for society to spend more time on schooling and preparation and less time on production. It is therefore reasonable to extend the period of training, to allow for study of unintended effects and yet have production sufficient to maintain a standard of living acceptable from the viewpoint of pollution and of other unintended effects.

For a number of reasons it is necessary to carry out part of this new education outside traditional academic life. I do not refer here to free universities. Rather, I am recalling efforts — both in my own country and in other countries in whose schools I have lectured — to bring to engineering students the concepts of responsibility, even where there are no formal courses on technology and society.

At Columbia University, I have held for years regular monthly meetings on these matters. Students signed up, although no academic credit was given. For each meeting one to three faculty members were invited to share their concern for broadening the concept of engineering or to recount life experiences which illustrated the widened views of engineering responsibility. For example, one colleague spoke about his lasting remorse for having refused to give expert testimony for the legal defense of two young black men. They were accused of murder by throwing a stone from a roof. He said that one look at the scene convinced him that no human being could throw a stone from the roof where these young people were apprehended to the place of killing. Yet, he was reluctant to take the time which would be required to attend the trial, because he did not want to give up his 'technical work'. Now he knew that he had done wrong by refusing to testify. A second colleague who had participated in the so-called Manhattan Project (developing the nuclear bomb) had concluded that nuclear armaments are wrong and he refused later work in this direction. However, being a mathematician and civil engineer, he accepted work for an organization in South Africa. He is opposed to the apartheid policy of this country but rationalizes that his case was a reasonable exception to the boycott which he otherwise favored: his product would outlive the unjust government and rules of apartheid in South Africa. The students were quick to challenge him on the grounds of self-deception.

Other schools have much more formal monthly non-credit-granting seminars on Technology and Society. I found one particularly interesting seminar at Stanford University in California, where even those monthly seminars arranged by the School of Engineering are open to all students from non-engineering schools. After my hour-long presentation came an hour of discussion. The students who had signed up came in groups of ten or twelve to faculty homes. Over dinner they continued the discussion in small circles, and heard other views than those of the speaker.

Difficulties will arise out of the newness of the concepts. How can we tell students at colleges that engineering must include the assessment of unintended effects, including pollution, when their employers may not accept this viewpoint and expect the engineers to carry out specific assignments disregarding unintended effects? This difficulty is very real and, as I have observed, occurs both in capitalist and in socialist countries. It demands education of engineers and of decision-makers, be they government or private employers. I believe that changing attitudes of decision-makers must be approached differently in the several countries because of differing social conditions. Yet, I know that international cooperation between those forces in the different countries who work for such change is essential — both for an exchange of experience and because we can strengthen each other.

Now I turn to the situation in my own country. Change *is* being attempted by pressure on the legislators — both directly (on the part of scientists and engineers) and through creating public pressure. The most outstanding exponent of this effort is Ralph Nader.

His approach and that of the numerous environmental-protection groups is mostly an attack on specific evils. Legislation has grown and is growing out of these individual cases. In the environmental field, important cases in point are the numerous requests of power companies to build nuclear power plants. Major efforts of a remarkably educational nature are aimed at convincing decision-makers that there is not enough known about the hazards of nuclear power generation, in the form of low-level radiation from effluents in normal operation, from transportation and disposal of nuclear waste, and dangers of major accidents. Consequently, the protesting groups urge that nuclear power plant should not be built near densely populated areas. Through the news media, public debate, and hearings the word spreads: much of America is densely populated: power demands are greatest in urban areas; obviously it is unfair that people living in rural sections should be burdened with radiation hazards incurred to benefit others who may live far away.

Engineering societies can play a major role in this re-education. However, this role is complicated because leadership in engineering societies comes necessarily from successful engineers, many so engrossed in traditional engineering that they have neither time nor interest in new views. Yet, at least one society in the USA shows a marked change. The American Society of Mechanical Engineers has a standing committee on technology and society.

This group works toward deepening and widening engineering education, bringing relevant programs to the society meetings and interpreting the new view to the whole membership. Most recently, the committee has proposed a new code of ethics for the profession.

In other societies there is pressure from younger colleagues to provide similar changes. This pressure takes in part the form of demonstrations on the floor of the meeting or of parallel sessions, where the societies refuse to take up the issues in the official program. The battle to make technology and science truly the servants of man instead of his masters is far from won, but, with gratitude, I report progress!

Now, it would seem to me to be very helpful if committees or divisions dealing with these issues in the several engineering and scientific societies in the different countries could cooperate. Here is a job in the organization of which the Society for Social Responsibility in Science should be willing to help.

As we are speaking of education and pollution we must never forget that a significant reduction of pollution means a genuine change in life style for everyone. In the USA and other technologically developed countries, transportation by private automobile increases. Even as we succeed in making cars with *less* pollution, it is beyond hope to have cars *without* pollution. Moreover, cars are made of materials, the production of which pollutes. Finally, the vast network of highways devours fields, woods, and farms that are desperately needed by man. So we must establish adequate public transport. In the same vein, the use of airplanes for short distances is 'woeful waste.' Other examples of change in the life style necessary to reduce and minimize pollution are too numerous to mention, and can be achieved only if the population at large is made acutely aware of the dangers which we face with our present life style.

The concept, often presented, that 'technology creates pollution', and, therefore, that 'technology can eliminate pollution' is false. Pollution will be reduced only when technologically developed nations limit, and probably even reduce, their per capita consumption. This requires education with zeal for people preparing for any kind of works, people being educated to take their part in public decision making; all must understand this relation of pollution and life style and of technology and society.

We know that minimal attention is now being paid to this terribly important matter in general higher education. We are confronted here with the desperate need to educate. At the same time, we must unfalteringly search for the right answers. As scientists we know that the oft suggested notion of abandoning technology is folly. Some sort of happy balance must be found between abandoning technology and its unlimited and uncontrolled growth. But to educate toward changes, the extent of which are as yet unknown, poses fearsome challenges. Yet without such education *now,* the future for the family of man is dark indeed.

I am convinced that education for goals described here must start in younger years and cannot be limited to persons headed for colleges and universities. Like some professors in universities, many teachers in our high schools are amazingly unaware of the problems.

My Quaker Meeting organized, in the small community, an informal discussing group of scientists and engineers, considering the field of 'man and technology' and the steps toward change. After about ten sessions, we engaged in an experiment. We invited twelve highschool students to an all-Saturday meeting. Around a fire we shared with them our hopes and fears regarding the use and misuse of technology. The following Saturday they returned to us, telling us first of their agreements and disagreements; second, how they thought other students might be reached and taught. A skillful teacher and several industrial engineers served as assistants. The students developed a fast-moving imaginative program, using films, music, dance, and drama, which was extremely provocative. Over twelve hundred students have seen this program and more invitations are now being received from other schools.

The message of their presentation and my message are the same: our world cannot continue without technology. But we dare not permit uncontrolled technology to destroy man. Therefore, 'Let us work while it is yet day.'

L. EMMELIN

An Environmental Studies Program

The Environmental Studies Program of the University of Lund is composed of all the faculties of the university, including the Faculty of Divinity. The program is concerned with raising the level of knowledge in order to make it possible for the community to receive and react intelligently to the warnings of scientists concerned with environmental problems.

Our belief that this education is urgently needed is strengthened by a small investigation, made by some of my students, which showed very clearly the lack of necessary background knowledge both by the general public and university students. The question simply is: How can one learn to understand what genetic damage means when one does not know what radiation or chromosomes are or what the former does to the latter and how this affects the lives of human beings? This lack of knowledge is particularly evident in Social Science and Engineering students — all of whom are most urgently in need of understanding what concerned scientists say.

It would seem logical to start by discussing what is to be done about education in schools. We at the University of Lund, however, decided that the trend in the Swedish school system at the present time is so reactionary, with respect to background knowledge, that we had better concentrate on teaching teachers who will teach in the school system when the pendulum swings back to a school curriculum that will provide good factual background knowledge for understanding these complex questions.

Two facts about the Swedish university system need to be emphasized. First, it is different from the US system in this respect: students take only one course at a time in Sweden; they usually take three to five subjects (three to five courses) to receive their baccalaureate degrees. This poses the problem that we have no common basis, not even within a group, to define students as science students or medical students, for example. There is no common basis of logic, semantics, etc. on which to build and which could serve as a common frame of reference for different students. Second, the form and the content of teaching are controlled quite rigidly on a national level, by the Office of the Chancellor of the Swedish Universities. This curtails local initiative. In fact, as far as I know, the courses taught in environmental science at Lund and other Swedish universities are the only example (in ten years) of any university making any drastic change in the teaching, or of any drastically new change emanating from a university — i.e., from the level that actually should be the moving force in changing teaching.

I know some people are going to dispute that, but I consider the bureaucracy that surrounds Swedish teaching as something that has not been experienced in any other country. The emphasis of the bureaucracy is very narrowly utilitarian. And this trend is strengthened by our present unemployment situation, which affects social scientists very seriously. At the moment, the major criterion for education in Sweden is job-orientation rather than orientation toward helping students to be good citizens or scientists. This is a fact of life that has to a great extent — perhaps too great — extent governed our actions. And this fact has inclined us to believe that we must emphasize including environmental considerations into all existing subjects, rather than try to create new courses.

There are other reasons for incorporating environmental studies into existing courses. Purely quantitively speaking, we could expect 22,000 students to come to our special courses. But it is also logical to interject environmental considerations into each subject. Unfortunately, introducing environmental considerations into all subjects is a slow process that is, in many ways, governed by the 'law' that scientists do not change their minds, but that eventually they do die.

As an interim measure, we offer an introductory course of half a term that is spread out over a whole term of evening classes. The course covers the major areas, splitting them up in the way Dr. Paschkis mentions (see pages 249–255). It includes an attempt at introducing the basic facts about ecology and discussing what we call 'Society and Environment.' This can be illustrated below.

There are two interesting points about this program. First, we admit all sorts of students. We have 300 students a year of whom about a hundred come from outside the university. We believe that they profit from coming to us. We also know that their presence has a very good educational effect on our teachers and lecturers, because they ask very sophisticated questions, far more sophisticated than questions normally asked by university students. For example, the town planning director of the city of Perth (Australia) was among our students. But we get more ordinary types of people too. Second, we organize students into groups of ten to do projects. This project work is directly action oriented and related to problems that we see around us. We believe that we do some good work and that we have contributed some of many small factors to the move toward establishing a traffic-free zone in the center of our city, for example. We try to have such practical projects for our students to work on. By having students from outside the university join our courses, we also help to educate our own students in project work; outside students bring practical knowledge.

Environmental Studies Program

Ten-point course	Forty-point course		
Introduction (2 points)	Introduction Chemistry	Introduction Law	(6)
Air Pollution and Noise (2)	Physiology	Economics	
Water Pollution (2)	Introduction Physical Geography	Introduction Human Geography	(6)
Conservation and Waste Management (2)	Ecology I	Planning	
Society and the Environment (2)	Ecology II (6)		
	Process Technology (6)		
	Natural Resource Management (8)		
	Environmental Health (8)		
	Planning Processes (5)		

Most of the other work we do is related to the problem of how to educate the needed experts in broader concerns. We do not attempt to 'train' anyone to be an expert, of course. In fact, we have taken a rather unpopular stand on this issue. Although we acknowledge the need for real expertise, such as that provided in the training of scientists, we also believe in the need for engineers, public health and other administrators, officials, and citizens to have some sound knowledge of environmental problems. This is not to say that we believe in the concept of an environmental 'general practitioner', or an 'environmentalist'. Trying to train such a man, we are inclined to think, would result in a quack. We do believe, however, that a lot of people, particularly those in public and business administration, need training that helps them recognize and predict problems themselves.

The students we admit into this course have to have a fairly good background, almost to the point of obtaining a degree in Engineering, Law, or similar fields. The course is arranged into two introductory parts in order to minimize the differences in the students' background knowledge. We recognize, generally, one group of Natural Science students and one group of students from the Social Sciences, Law, and so on. All need to complement their knowledge in the social sciences more than the social science to the belief of most natural scientists.

A few words about our experiences. First of all, it is easier — and it is much more fun — to teach these types of courses than most traditional academic teachers thought it would be. It is good education for the professors. Second, discussions have strengthened our belief that good training in some subjects must precede environmental education. Our belief is definitely that students should have a program. They should also have training that prepares them for a job, even if the environmental education does not do so. Because there is no such profession as that of environmentalist, there are no jobs for environmentalists.

Lastly — and privately — I have begun to doubt the level of sophistication of the subject 'Ecology.' We see that the knowledge of ecology, and the standard of teaching ecology, is not compatible with the standard of teaching and the standard of knowledge that our engineering students or our law students have. This might be a trivial problem. It might be simply that we do not have good teachers. But it seems a real problem and a danger. In Sweden, at least, ecologists have launched themselves, on a large scale, as the only 'real' saviors of humanity. And the engineering students, particularly, are beginning to doubt that this is so and to think that ecology is not all that important, even that it is a bit of a humbug.

By drawing largely on the very meager resources of the ecological departments, we are diverting effort and work away from basic research, which is the only way to raise the level of sophistication of ecology.

WOUTER VAN RAAMSDONK

Promoting the Sense of Social Responsibility in Dutch Education

This paper is intended as a contribution to the discussion which has recently sprung up over the problem of social responsibility of scientific workers. This is an issue which holds the attention of many students and teachers; articles on it appear in various newspapers and magazines; and it was one of the main themes of a symposium that was held recently at the Technological University of Twente (the Netherlands) by the World Federation of Scientific Workers.

Ever fewer scientific workers will deny — at least in public — that they have a responsibility toward society. Nearly all discussions, reports, articles, and papers on the subject of responsibility toward society are based on the presupposition that this responsibility can in fact be possible; too often social responsibility is presented as being a problem only for the peace of mind of the individual researcher.

In this paper I will give a survey of the promotion of the sense of social responsibility at universities in the Netherlands. I have on the whole mentioned only projects and courses that were set up at my own university (Free University, Amsterdam) to illustrate this survey, but, the Netherlands being a small country with only a few universities, which are, moreover, very similar both in academic level and mentality, these examples can be taken as representative of the situation in the Netherlands as a whole. I will then consider consequences this might have. In the light of these considerations we will be able to make some more general observations.

The institutions of higher education referred to hereafter should be taken to mean those institutions mentioned in the Dutch Higher Education Act, viz. the ordinary universities and the technological, economic, and agricultural universities.

In this paper I understand the promotion of the sense of social responsibility as the instruction to make the relationship between science and society a subject for discussion in teaching as well as research, and

260

to provide room within the curriculum for students and teachers to evaluate this relationship and to organize social action that follows from it. This I would like to call the minimal interpretation. There are also people in favor of a much wider interpretation. of the social responsibility of universities. They maintain that the university *as an institution* must give evidence of its social responsibility by committing itself to a standpoint and by campaigning for social reform.

Personally, I am inclined to support the first, minimal interpretation: in a society which lacks a generally recognized set of norms and values the university as an institution cannot be expected to commit itself to a standpoint and to set up campaigns. After all, all institutions for higher education in the Netherlands are financed from public resources.

The aims of higher education have been set out in the Dutch Higher Education Act. They are: research; teaching; promotion of the sense of social responsibility and of understanding of the interrelation of all sciences.

The latter two aims do at some points coincide, but should, nevertheless, be considered separately. In this paper, I will confine myself to the third aim: promotion of the sense of social responsibility.

During the parliamentary debate of the Higher Education Act, in 1960 and afterwards, the article dealing with promotion of the sense of social responsibility was strongly criticized. One group thought that inclusion of this particular article was superfluous, while another wanted a much more detailed text; a third group maintained that science should be used exclusively for science's sake. But, the latter group found little support for this idea.

The term 'promotion of the sense of social responsibility' only began to acquire a definite meaning some years later. Influenced by events in Germany *(Kritische Universität),* students in Holland started a movement which aimed at using science to create a new social structure. Science should, once more, play an emancipating role, i.e., champion the interests of underlying groups in society.

The appeal the students made on the article from the Higher Education Act dealing with promotion of the sense of social responsibility enabled the university councils to meet at least some of their demands. During the years 1967–68, socio-educational councils were set up at most Dutch universities. These committees, which consisted of students and teachers, were given the task of looking after the social aspects of the students' training. (The traditional student federations which before pretended to perform this task, have lost a great deal of influence.)

Besides socio-educational councils there are two more university institutions that are engaged in promoting the sense of social responsibility. The first are the philosophy departments, which in the Netherlands are traditionally independent of all other departments and form a separate 'inter' department. The second is the committee for the *studium generale*, which is in charge of a programme in which 'the unity of all sciences' is the central theme. In practice, this programme entails a series of lectures on one particular subject, given by people from different branches of science, who will then all approach the subject from their particular point of view. These 'studium-generale' courses are always on a voluntary basis, and the students cannot get credits for them.

There are also activities within the scope of the departments' curricula aimed at making students and teachers more aware of their responsibility towards society. Some examples from my own university are: an education project in the department of social sciences about migration-labour; a series of lectures on 'science and society' in the departments of mathematics, physics, chemistry and biology, and more recently an environmental project which was also set up in the departments of mathematics, physics, chemistry, and biology, but which hopes to cooperate with the departments of law, social geography, economy, etc. These are only some examples; other universities in the Netherlands have started similar projects.

The activities mentioned in this paragraph differ from those mentioned above, in that they form part of the students' official curriculum and in that the students can get credits for them.

At all Dutch universities there are, moreover, more or less informal groups of students, who are working on the problem of responsibility towards society. These groups operate, however, outside the universities, and are often connected with professional, non-university organizations, such as *Verbond van Wetenschappelijke Onderzoekers* (Federation of Scientific Researchers, the Dutch affiliation of the World Federation of Scientific Workers); *Bond van Wetenschappelijke Arbeiders* (Union of Scientific Workers); *Vereniging voor Medische Polemologie* (Association for Peace Research in Medicine); *Kritische leraren* (Association or Radical Teachers), etc.

I have tried to indicate the different levels within the university structure where efforts are being made to define the concept social responsibility of the scientist.

Regarding the content of problems involved there are three main areas where social responsibility plays a definite role: control of the environment; the third world; peace research; and some miscellaneous problems.

Although pollution of the environment has been going on for quite a number of years now, the university world has only recently given evidence of its concern in this matter.

In the Netherlands, biologists were the first to take up the matter

when they took part in a campaign to prevent the setting up of a carbon disulphide plant in Amsterdam. Pollution is a problem which stirs people up easily, and therefore there are also many people not connected with the university sitting on action committees. The major issues in the field of environmental control are at present:
— The establishment of petrochemical industries (in Amsterdam: application rejected; in Antwerpen, in Belgium, just across the border with Holland: application recently approved).
— Aircraft noise in the residental areas near Schipol Amsterdam Airport (a problem which has not yet been solved).
— The construction of a nuclear power station near Leyden, which would cause the water temperature in the area to rise and would severely pollute the environment.
— The chemical industries at the industrial estate near Rotterdam (along the *Waterweg,* Rotterdam's route to the open sea).

Not only more or less organized groups of people (scientists and non-scientists) living in these areas are active in the battle against pollution, but also the universities on this point. Several universities are in the process of setting up 'environmental control' courses, where the pollution problem will be studied from various disciplines, such as biology, social-geography, economics, sociology, and law.

As in the case of environmental control, the problems concerning the Third World have been mainly brought out into the open by unofficial action committees. But some universities (Rotterdam, Free University at Amsterdam) have created a Chair of Development Economics within the department of economics. Especially, Professor Jan Tinbergen's Institute at the Economic University of Rotterdam is world-famous. But the problems of the development of the Third World contain more than just economic aspects. The Free University of Amsterdam is studying the possibility of a course on 'development problems', where — as in the case of environmental control — the subject will be approached from many different sides: economics, cultural anthropology, international law. etc. The aim is to make it a free choice subject for undergraduates.

Contrary to the two above-mentioned issues, the development of peace research was started at an institutional level, and only later taken into the action committee sphere. Professor B.V.A. Röling's Institute for Peace Research at Groningen University was set up ten years ago. Efforts were made to establish a similar institute at the Free University, but, in spite of active student campaigning, these have failed. The department of social sciences, however, has appointed a lecturer who will be specially in charge of peace research problems. Nijmegen University has already established a sub-department of peace research, within the department of social sciences. Some time ago the Dutch government has set up an institute for peace problems *(Nederlands Instituut voor Vredesvraagstukken)*, which will only carry out research and advise the government. The usefulness of this institute depends on how the results of its research will be made public, and on whether it will serve as more than an advisory body for the government.

In Amsterdam, Utrecht, and Leyden, moreover, there are groups of students — sometimes with teachers — engaged in studying peace research. Occasionally this can be done as part of the syllabus of the subject studied (sociology, Leyden). Some years ago, Groningen University organized a short series of lectures on peace research; the Technological University Twente has done the same during the past year.

Under the term 'miscellaneous' a number of different projects concerning the relation between science and society have been collected. The university socio-educational councils are concerned explicitly with the relationship science-society, and an important part of their task consists of guiding and aiding project groups etc. studying this relationship. During the past year, the socio-educational council at the Free University started a project 'computerisation of society'. Last year the 'studium generale' committee at the Free University organized a series of lectures on the changeability of man, where the relationship man–science–society was one of the central themes. The activities that go on at departmental level are more directly linked up with the situation people concerned are in. The department of mathematics, physics, chemistry, and biology at the Free University has for instance organized a series 'science and society', the aim of which is to bring the students to a better understanding of society and of the role their own particular field of science plays in it. It is hoped, with good prospects, that this course can be continued as a free-choice subject for undergraduates. The universities of Leyden and Utrecht have set up similar courses, in particular for chemistry students.

In the medical world most of the activities are still on an informal level. Groups of physicians and medical students have questioned the often 'accommodating' nature of the role of the physician. They wonder for instance whether it is right to prescribe tranquilizers to victims of the (in Holland still pressing) housing shortage without trying to remedy the housing shortage and the social structures that produce it.

One of the most important problems in the educational field is to what extent education must or can conform to the established patterns of society, or whether education should actively work towards social reform. Not only groups of radical teachers are trying to find an answer to these questions, but the problem is also being vividly discussed by students in pedagogy.

264

The prominent professional organizations that are studying the relation science—society are the V.W.O. and the B.W.A. the *Bond van Wetenschappelijke Onderzoekers*, the Dutch affiliation of the W.F.S.W., is setting up a project on 'ideology of the Western World', in which project groups and congresses will endeavour to establish the — usually hidden — ideology behind the application of sciences. An introductory congress on this subject and one dealing with the ideology of economic structures have been held in recent past. The ideology of economic structures is also the study-object of a group of students led by Professor van de Klundert at the Economic University in Tilburg. In this case, the subject of study is part of the syllabus.

From the examples mentioned above it will be evident that a considerable number of scientific workers in the Netherlands are making a serious effort to promote the sense of social responsibility and actually practise it.

It will also be clear from this survey that, at present, promotion of the sense of social responsibility is by no means a first priority of Dutch Higher Education. There is, it is true, some action in this field at an institutional level, but this is cerainly no general movement. The activities at non-institutional levels are on the whole rather fragmentary and marginal; that is one of the reasons why they are unable to effectively influence teaching and research at the universities. This rather gloomy picture I have painted of the situation may, however, not be entirely fair. Perhaps it is better to see things in a historical perspective and trace the developments that have taken place since 1945.

For every individual it is a different issue that will make him aware of how society works and of the remedies he himself can suggest and work on.

In the years following the Second War this issue was the use of nuclear weapons, during the Sixties it was the war in Vietnam.

The effects these two issues produced — a growing awareness of the responsibilities scientists have as regards the consequences of their work — did not differ fundamentally. Today, scientists may not take as long as before to become aware of their responsibility: After the Second World War, people were shocked in their ethical values and felt, as human beings, morally responsible 'for the consequences to humanity of professional activity', to borrow a phrase from the SSRS Newsletter. Then a process of political awareness was initiated. During the Sixties the Vietnam war had the same effect; it took less time, maybe because there was less interference with this process: those who adopted a critical attitude were less harassed by insinuations than those who expressed criticism in the days of the cold war. The latter group could, in the last instance, only retreat to its final barrier, that of moral indignation. With the change, however, from a way of thinking in terms of personal, moral indignation to one in terms of the social structures in which science functions, in other words a change from micro- to macro-ethical values, the question rises whether responsibility is possible.

The responsibility of the scientist as such is based on his scientific expertise, This expertise contains three elements:

(a) knowledge of a particular field of science;
(b) knowledge of the interrelation between this field and certain others; and
(c) knowledge of the interrelation between this field of science and all other sciences and the society in which it functions.

It is the third aspect of the scientist's expertise that is important when we talk about his social responsibility; but it is also the least developed aspect of his scientific expertise.

If science means that mankind behaves rationally, i.e., its actions are based on reason, and in this way liberates itself from the forces by which it is ruled, then the development of the third kind of expertise is an intrinsic part of science and rightly belongs to the responsibility of the scientist.

In his forthcoming book, *The Academic Nowhere,* Professor J. F. Staal distinguishes four functions of the university:

— the pure function, which centers in teaching and research leading to pure knowledge sought for its own sake
— the service function: the service of the university due to society; without the ideal of the service function universities would not have faculties of law, medicine, economics and the like
— the critical function; the duty to evaluate critically society as a whole. This critical function was quite widespread in the past and has been generally feared and disliked by those in authority
— the leadership function: to provide intellectual and moral leadership.

According to Staal the four functions should be balanced carefully. Once it is accepted that universities should exercise the service function which is regarded (and rightly so) as compatible with the pure function, it becomes impossible to maintain that the critical and leadership functions are incompatible with the pure function and that they therefore should be avoided.

Just as science enables us to give more adequate statements and descriptions and better justifications, it also enables us to produce better evaluations. It is one of the features of the service function of scholarship that scientists enable others to pass value statements. It would be strange indeed if scientists themselves were prevented from making such evaluations.

If this opinion about the four functions of the university is right — and I think it is — then the development of knowledge concerning the relationship between science and society should belong to all four functions. This means that research must be carried out, as objectively as possible, i.e., subject to continuous criticisms and rectifications, into how certain fields of science influence social developments, and how society influences the development of certain fields of science. It also means that scientists must use the knowledge they have thus acquired to promote favorable developments and check unfavorable ones. This should be done by committing oneself to a political standpoint, based upon the specific knowledge of the researcher, or the professional organizations, or even, but that is still farther away, the university.*

A beginning will have to be made by collecting more information about the relation science—society. The Dutch universities have already started, although it is no more than a modest beginning.

It is significant that nearly all courses I have mentioned are not disciplined-oriented, but problem-oriented. Activities are, therefore, of an interdisciplinary nature: only through cooperation with sociologists, psychologists, jurists, economists, historians, etc. can we come to an understanding of the complex problems of war and peace.

Environmental control is not a matter only for economists and biologists; it also concerns planning experts, sociologists, administration experts, etc.

In my opinion these activities should become an obligatory part of the study of certain subjects:
— because of the principle that this sort of activities is an essential part of the university's task;
— for the practical reason the tendency towards shortening university programs, which is imminent in the Netherlands as elsewhere, leaves students and teachers with less and less free time.

In my opinion, this is at present the only possibility to get a grip on teaching and research. The idea has evident disadvantages: this new part of the study could easily be used as an excuse to ignore the subject in other parts of the study, which remain non-committal. This way of thinking might be expected from both teachers and students.

* In Staal's opinion, the national institutions will not be capable of doing this; he argues in favor of internationalization of the university.

I think, however, that this danger can be largely removed by committing the students to such a project from the very beginning and making their cooperation an essential part of it. Responsibility, after all, can be acquired by undertaking it. The courses must be organised in such a way that the students feel responsible for the proper functioning of the group and for the results they achieve. There should not be a teacher in the traditional role: he must be adviser and administrator to the group. The group as a whole, for instance, is to choose the subject of the course. When in this way practical training in the sense of social responsibility is given during a course on social responsibility, it is likely that the training will not remain an isolated spot within study and practice, but will have the effect of an oil-stain. Only then we speak of a change in mentality in university research and teaching.

The very modest beginnings that have now been made at Dutch universities can be appreciated positively only if they appear to be a phase in a development toward a change in mentality within the university, a change aimed at making teaching and research a means to create a just society.

Summary

This paper gives a short survey of the situation in the field of promoting the sense of social responsibility at universities in the Netherlands.

The Dutch Higher Education Act explicitly stipulates the promotion of the sense of social responsibility as one of the aims of education. Though the wording of this article of the law can be interpreted in various ways, it is workable as means to pin down the university community to its responsibility toward society.

First of all it is explained in this paper on what levels the promotion of the sense of social responsibility takes place: on central university level, on faculty/department level, on non-institutional levels within the university community, and finally on the level of the professional organizations. Most of the activities take place on the latter two levels, in the former two cases it is often more a matter of theory than of practice.

Furthermore the activities are distinguished regarding their intrinsic aspects: distinction is made between activities in the field of peace research, of third world problems, of environment control and various activities like medical ethics etc.

As there is a growing tendency in the Netherlands to shorten university programs and to organize them more strictly, it is high time to design interdisciplinary courses aimed at a number of cognate studies in which the relationship science-society is central and which are part of the normal study-program for all students. Characteristic of such courses should be that they are problem-oriented, not discipline-oriented. The disadvantages of such a concept are evident: a like part of the study can function as alibi and leave the rest of the study-program in the often still usual non-committal sphere. And yet, in my opinion this is the only possibility — at least at this moment — to force an entry in official study-programs.

Some of the Dutch universities did make a beginning with it. The next phase should be to bring about a change of mentality in the entire university-education from these nuclei, in order to make education and research subservient to bringing about a just society.

Toward Solutions

HARRY BRIGGS

Scientific Leadership and the Price System

The Menton Statement presented to Secretary General U Thant at a special ceremony of the United Nations on May 11, 1971, strongly underscores the dangerous environmental conditions threatening life on this planet.

For years now, scientists and concerned individuals have been drawing a frightening picture of a world suffering from overpopulation, misuse of energy and technology resulting in world-wide pollution, existing wars and the threat of world war. The purpose of this paper is to identify basic causes and, more important, to suggest possible solutions, rather than to restate and redefine the foregoing problems.

A principal reason for the problems at hand is the lack of planning for the transition of societies from economics based on hand tools and human toil to technological social operations. Since governments in all their forms, past and present, have been designed to control *people,* the age of technology leaves them totally unprepared to cope with the problem of controlling *things.* The evident fact that the more technologically advanced nations are not managing their newly found power intelligently is dangerous enough in itself. But worse yet is the fact that emerging nations have no example to follow in their development, and thus may make the same mistakes more advanced nations are making. This condition tends to leave underdeveloped nations at the mercy both of the advanced and emerging nations in a sort of cannibalistic resource exploitation that would destroy any chance of a future for the underdeveloped.

While world conditions present a dismal enough picture, I would draw attention to a new, decisive factor in the American economy: money-debt problems. They have upset its internal economy and the international monetary system, and will doubtless have a great bearing on future U.S. relations with the rest of the world and could very well be a powerful influence in the shape of things to come.

The predisposing factor in the increasingly shaky U.S. monetary system is the enormous debt — public, corporate and private — that has accumulated over the years, especially since 1946.[1] We are approximately $2 trillion in debt. Attempts to ease the money problem to service this huge debt have only resulted in more inflation.

The precipitating cause for current US money problems is a steadily accelerating inflation. For years the inflation rate of US money was one to one-and-one-half per cent per year, with the exception of the Korean War period, when it moved higher. The present inflation correlates directly to the expansion of the war in Indochina. It began to accelerate in 1965 and has moved rapidly higher each year, until the inflation rate now is in excess of seven per cent per year.

No one thought the Indochina War would cost so much — more than $125 billion — or last so long. Now, too late to save its financial system, US policy-makers attempt to retreat from the war. But the damage has been done. A chronic money crisis has been created by a failure to provide for the war cost, and the most corrosive kind of inflation besets the US dollar, which since 1960 has lost 27 per cent of its purchasing power, according to *The Wall Street Journal* of February 8, 1971.

Inflated dollars, that have been pouring into European central banks, are inflating one European economy after the other. The efforts of US monetary authorities to control inflation have been ineffectual, which is reflected in the steadily climbing gold bullion price, now at the historic high — on August 4, 1971 it was $42.50 in London — as people and economies run from paper money, principally dollars.

The money problems in the US are causing business failures, strikes, personal bankruptcies and economic stagnation in general, problems tending to cause a physical disintegration of the US economy. World trade and balance of payments deficits are adding to the problem. Seattle, Washington, my home town and site of a major segment of the airframe industry, is a prime example. Some 70,000 workers have been laid off by the Boeing Company in only a little over two years because of declining business. More than 20,000 formerly affluent people in Seattle are actually facing starvation. There is no significant relief in sight for Seattle, and this appears to be spreading across the country. Money, as the USA has known it, is failing as a means of moving goods and services.

Given a continuation of present trends, some of us say the government should be forced to declare a moratorium on all money transactions, for, to resort to more fiat money manipulations can only lead to a further loss of confidence in the economy, and finally economic chaos. The present money crisis is not the first in US history; but it is by far, potentially, the most severe. And it appears to be insoluble.

274

The situation in which Americans find themselves calls for a new concept of social operations. North America has a relatively homogeneous population and a contiguous land mass that is becoming technologically integrated, a natural place to start a new pattern of social operations which might be an example to the rest of the world. Something entirely new must manifest itself — a new idea of social operations — for no system of politics or economics operating in the world today is capable of dealing with the magnitude and complexity of its problems.

To avoid the possibility of physical disintegration in North America, the first priority is to keep the wheels turning and the gears meshing, regardless of what happens in the worlds of money and politics. The basic functions of society are those that count in terms of human life. As it goes through the trauma of massive financial disintegration, North America must begin mobilizing its knowledge, skills, tools, and resources in the common interest of survival, in order to prevent industrial collapse.

Next on the priority list is development of designs for mass transit, rehousing, education, medical care, power transmission and waterway development. Designs for these major functions of society will materialize when there is a proper coordination of all social assets that really count in our matter-energy world.

It is fair to say, the people of the US and North America have performed some remarkable feats in their brief history. Seattle, under the direction of the Municipality of Metropolitan Seattle, implemented a comprehensive plan to clean up Lake Washington — approximately 18 miles long, 3 miles wide, and up to 240 feet deep — after this ravine-like lake was threatened with being choked by algae.

Within two years, between 1967 and 1969, when large essential portions of sewage diversion began operating effectively, Lake Washington substantially cleansed itself. Transparency of the water is now 10 to 11 feet, phosphorus is down to 22 parts per billion[2] and the lake is safe for swimmers and for other recreational use.

A larger example of comprehensive planning is the North American telephone system, which operates entirely by dialing, without any operator assistance, anywhere in USA or Canada. Besides carrying the telephone call load, the system carries radio and television signals and other types of communication. The system is proof that problems may be approached and solved on a continental scale with brilliant and durable success.

Regardless of the controversy and criticism involving the space research programs — even though they in themselves are inordinately costly for any one nation to bear and seems to produce results not commensurate with the effort — they prove that man *can* design and operate intricate systems to support life and they have reinforced a realization that Planet Earth is all there is in the foreseeable future to support human life.

Within the system of life on this planet must come balance, if it is to continue to support life. The idea of balance, that we cannot take more than we put back without paying in human suffering, must extend into every individual's life and daily activity. It is therefore proposed that a new calendar be developed in order for all functions of society to operate on a balanced load, thus making maximum use of existing facilities, on a 24-hour basis, wherever possible, seven days per week, and year-round. Simply distributing the load more evenly around the clock on industrial, transportation, power educational, medical and recreational facilities would greatly diminish energy use, while achieving a much more efficient result in terms of the production of goods and services. It is worth noting that the go-stop syndrome of most business activity is an extremely costly feature to society that cannot be afforded in the future.

Reference is again made to the money-debt problems of the US and the problems this is posing for the world. One is forced to conclude that a new accounting system, tying in the consumer directly in the production of goods and services, is the answer to an orderly movement of goods and services and the only way to avoid a chaotic social condition in North American civilization.

It is suggested here that a Distribution Card that would measure energy flow in production and use of goods and services, be implemented in North America as the consumer instrument. The data-processing equipment already exists to process information on rates of use of all goods and services. It is only a matter of issuing the Distribution Card. The Card would be issued as a matter of birthright and would have no connection with what sort of service the holder performs, thus avoiding the complicated problem of trying to determine what each individual is 'worth'. One would anticipate that, as a matter of economy and efficiency, only the highest-quality products possible would be produced for all individuals. We truly have reached the point where either we produce the best for all, or finally there will be nothing worth having for anyone. The latter possibility appears to be only a few years away, if present trends continue.

As the present system — properly called a price system — fails because of its own defects, it is recommended that a Grand Alliance of functional people of industry, educators, students and the general public be formed to fill the leadership vacuum developing now. It is stressed here that such an Alliance must be orderly and that it proceed with these objectives.

276

1. Lay plans immediately to keep all industry essential to human health functioning.
2. Lay plans to keep all activities that cause pollution to the barest minimum.
3. Proceed with the development of new designs to provide facilities, goods and services mentioned above.
4. Undertake immediately the widest possible public education program to acquaint the public with the perils, problems and available answers, stressing at all times the positive, problem-solving approach. Defeatism is a major enemy of human survival.

The foregoing sounds enormously ambitious. Yet, the problem size dictates the solution size. It has been said that 'we must do the impossible or face the unthinkable.' It is asserted here that North America is truly representative of all ethnic groups and backgrounds in the world, which, we think, gives that continent a peculiar strength and special responsibility to lead.

In addition, North America has enormous physical technological power to aid other nations of the world in a fight to save this planet. The leadership to do this consists of 1,747,800 scientists, engineers and technicians;[3] 2,994,000 teachers at all grade levels;[4] 338,379 physicians, surgeons and osteopaths,[5] and millions more of highly skilled workers and managers in every conceivable kind of human endeavor.

We trust that the competent and intelligent people of this world will encourage North America to consolidate into one operating unit, using the method of science (without worshipping science), to accomplish the great changes that must take place in the common world interest.

CITED REFERENCES
1. *Congressional Record,* June 29, 1970, p. S10196.
2. Statement by Robert P. Hillis, Municipality of Metropolitan Seattle, July 28, 1971.
3. U.S. Statistical Abstract, 1970, p. 523.
4. U.S. Statistical Abstract, 1970, pp. 123, 127.
5. U.S. Statistical Abstract, 1970, p. 65.

L. BICKWIT, JR., AND B. B. BROWNLEE

Chemical Pollution Control in the USA

The US Congress is currently undergoing a soul-searching review of many regulatory mechanisms to control pollution. Late in 1970 legislative amendments were passed to strengthen the control of air pollution, and Congress is currently considering further amendments to our pesticide and water pollution laws. In addition, a new proposal has been introduced, at the request of the Administration, to control other toxic substances that are not currently regulated. In conjunction with this proposal, Congress is deeply concerned with two basic questions: How can we rationalize our various regulatory schemes in the field of chemical pollution into a consistent scheme? What should be the components of that scheme?

As similar questions undoubtedly must be faced by other nations, it seems appropriate to discuss these questions. As our own expertise extends only to the US statutory framework, we shall, therefore, confine our remarks accordingly. We shall rely heavily on discussions that took place at a recent seminar with participants from government, industry, and the scientific community.

The noted researcher, Dr. Samuel Epstein, a participant in the seminar, stated in his presentation: 'Federal regulatory agencies suffer from certain limitations and constraints. Jurisdiction over products and their applications are fragmented among the sixty or so executive agencies concerned with environmental problems. Additionally, there are certain categories of products, notably industrial chemicals, for which there are no or minimal regulatory controls.' He further complained that 'legislative responsibility in environmental and related areas now extends to some twenty congressional committees.'

As Dr. Epstein suggested, the resulting complexity is unfortunate, to say the least. On the federal level, standards for exposure to chemical pollutants through the media of air and water are established by the Environmental Protection Agency, and tolerances for exposure through food are generally set by the Food and Drug Administration, while standards in the work place are the province of the Department of Labor. It is often alleged that there is insufficient coordination among these standard-setting agencies, so that when standards are set for one source of exposure, insufficient account is taken of likely exposure from other sources.

The problem is one not merely of too many agencies, but also of too many laws with inconsistent provisions. If dangerous mercuric compounds are used in pesticides, pre-clearance by regulatory authorities will be required before the products can be placed on the market. However, if similar compounds go into cosmetics, no such pre-clearance will be necessary. Moreover, industrial uses of mercury are subject to no direct regulation whatever, but are controlled rather through regulation of the effluent or emissions of the industrial plant. Numerous other examples can be cited of equally dangerous chemicals receiving disparate treatment under the numerous federal laws.

While less fragmentation and increased consistency would be highly desirable, it is unlikely that we shall see those objectives achieved in the near future. To produce one omnibus law dealing with all forms of chemical pollution would require a tour through each of the Congressional Committees Dr. Epstein referred to — a tour which no bill could ever hope to complete in the course of a given two-year Congress. The alternative is to amend each of the existing laws to make them as consistent as possible with one another, and to supplement them with new laws in areas where no regulation is yet required. While this is a worthy objective, it would take a great deal of time, if indeed it could ever be achieved. Again, similar sentiments would have to be found in the various committees as well as similar resistance to lobbyists opposed to additional controls — a highly unlikely prospect.

This route, however, certainly ought to be explored. Perhaps the best way to proceed would be for Congress to attempt to enact model legislation in areas such as industrial chemicals, where all agree that insufficient controls now exist. Pressure might then be applied to attempt to make existing statutes conform to that model, departing from it only in situations where differences in products truly justified departures. Requirements might be imposed upon existing agencies to coordinate standard setting and enforcement activities with each other. Additionally, further thought could be given to increased consolidation of these agencies. While there are obvious limits to the extent to which consolidation is desirable, the President's Reorganization Plan Number 3 of 1970, which established the Environmental Protection Agency, certainly seems a step in the right direction. If the Agency proves adequate to its numerous and rather awesome responsibilities, consideration ought to be given to further expanding its role in the field of chemical pollution.

Dr. Epstein made numerous welcome suggestions on the possible reformation of regulatory procedures for chemicals. Essentially they related to the following problems: When should a regulatory decision regarding the use of a given chemical be made? How should it be made so as to avoid 'constraints between industries and regulatory agencies'?

Regarding the timing of decisions, Dr. Epstein was persuasive in his view that these decisions should be made at the earliest possible time. Rather than cage the lions, it seems sensible to do all we can to control the lion population! The Food and Drug Administration now faces a mammoth task in inspecting fish and other foods for toxic materials that are, at least partially, due to industrial waste discharges. From the standpoint of both safety and economics, it would seem that we would be much better off had the chemicals that produced these discharges been controlled at an earlier stage. Without blaming anyone in particular for having allowed an unfortunate situation to develop, pre-clearance as a mechanism for pollution control would seem indicated after our experience with mercury and other toxic metals.

Accepting that decisions involving the manufacture and use of chemicals should be made at the earliest possible time, we are left with a central question of how to make those decisions as enlightened and objective as possible. Obviously, increased research on the efficacy and potential dangers of given chemicals is necessary. Less obviously, perhaps, is the need to eliminate the industry-agency 'constraints' to which Dr. Epstein referred. It is necessary to insure the accuracy of industry data, possibly by means of 'a disinterested advisory group or agency to act as an intermediary between manufacturers and commercial and other testing laboratories.'

Whereas Dr. Epstein's proposal is interesting, one needs to question whether it is strictly necessary. In terms of cost effectiveness, it might be an advantage to continue the practice of allowing industries to do their own testing, but to spot-check their results and to impose stiff penalties when a manufacturer's data are found to be inaccurate. This assumes, of course, that industry will be alerted to what tests are required and what results must be obtained in order to allow the chemical on the market. On this point, Dr. Epstein wisely underscored the need for promulgation of exact protocols to guide industry in its testing.

The question remains, however, as to how these protocols and, more importantly, the criteria for safety and efficacy for new and existing chemicals, are to be determined. It goes without saying that, even if the data presented are perfectly accurate, if the criteria industry is required

280

to meet are inadequate, we have achieved nothing. Moreover, it should be noted that, if constraints are present with regard to the presentation of data, they may also be present with respect to determination of the criteria which the data are required to satisfy.

As a solution Dr. Epstein suggested 'stronger scientific and legal representation of consumer, occupational, and environmental public-interest groups in all agencies concerned directly and indirectly in these areas' and consideration of 'the possible need for an independent consumer regulatory or protection agency.' This is similar to the proposal, introduced before Congress by Senator Abraham Ribicoff and Congressman Benjamin Rosenthal, to establish an independent consumer advocate. To aid in this consumer-advocacy function before agencies, Dr. Epstein further proposed increased public access to data relevant to agency determinations.

These suggestions we find highly desirable. Our only concern is that they may not be sufficient. If, in balancing benefits against risks in determining safety criteria, an agency is caused by industry pressure to concentrate on the benefits rather than on the risks, would increased consumer representation in the administrative agency really be likely to change the agency's bias? If an agency is disinclined to restrict the use of a chemical because the chairman of its congressional appropriations subcommittee opposes any such restriction, is a public hearing really going to reverse that disinclination? We know from experience that regulatory agencies frequently get friendly with those regulated and that they are influenced by the politics of the particular situation surrounding their actions. It will be the rare case in which consumer advocacy within the agency will be able to match these pressures. Moreover, if the consumer advocate is itself an agency of the US Government — as the Ribicoff-Rosenthal proposal would have it — it may be subject to the very same pressures as the agency before which it argues.

We would suggest, therefore, that something further is needed to insure that decisions of such great importance will be made objectively on the merits and in the absence of the traditional biases of regulatory agencies. There are several alternatives that should be examined.

1. Congress might take it upon itself to make pertinent decisions. Several people have advocated this approach by introducing bills to ban DDT, other chlorinated hydrocarbons, and several additional chemicals. We are critical of this method for two reasons. First, Congress is subject to many of the same pressures that are exerted upon the agencies. Second, the competence of Congress to reach intelligent decisions on highly technical matters must be questioned. Although decisions are ultimately political, risks and benefits must be balanced and the problems solved in terms of questions of values. But although we tradition-

ally look to the legislative branch of government for value judgments, it is inadvisable to do so in such cases, because members of the Congress would have to devote an enormous amount of time to studying the technical evidence in order to understand the true dimensions of the factors that must be weighed before reasonable decisions could be made. Lacking such understanding, members would inevitably make important decisions on the basis of uninformed constituent preference and without sufficient regard for the merits of the problem.

2. Congress might tighten the guidance given to administrative agencies in making these decisions. Congress could set down with greater specificity what considerations the agency involved is to weigh and in what proportions such considerations are to be balanced.

An example of this approach may be found in S. 808, a pesticide control bill introduced by Senator Philip A. Hart before Congress. The bill provides that a pesticide must be banned whenever there is a reasonable doubt as to its safety for humans or the environment and wherever a reasonable alternative exists about whose safety there are less serious doubts. Whereas this kind of guidance is desirable, there will inevitably be limits to its effectiveness. In the case of the Hart bill, the administration is still left to determine what is a 'reasonable' alternative. Thus, he is still required to do the rather delicate interest-balancing which is all too frequently influenced by irrelevant 'constraints.' Unless Congress is actually to make the decision — which we consider inadvisable — the agency will have to be given some room for such balancing. Although it might be possible to confine the problem with increased legislative guidance, it certainly cannot be eliminated by that route.

3. It has been suggested that the best way to eliminate the problem is to change the nature of the agencies themselves. New selection and funding processes, designed to increase the independence of the agencies from objectionable political pressures, may be needed.

As far as selection is concerned, it does seem that the traditional process is undesirable. Top administrators, who are initially chosen by the President, quickly become identified with 'the Administration.' Their resulting preoccupation with perpetuating the Administration is obviously an unhealthy factor. Although there is nothing objectionable in political responsiveness to public wants and needs, the responsiveness which this orientation generates is often to the wants and needs of campaign contributors.

Efforts to 'depoliticize' the selection process are thus very much to be applauded. One possible advance in this direction would be to give agency administrators increased tenure so that they could not be dismissed at will for unpopular, but meritorious, decisions. But when one attempts to progress beyond this point, the going gets tougher. Some argue that Congress itself ought to appoint agency administrators. Yet since Congress is also a political actor, it is hard to see how this would achieve the desired result.

Even if an adequate selection process could be devised, the problems of funding would remain. If agencies are to remain dependent upon the appropriations process, the possibility of pressure by a particular subcommittee chairman will continue to exist. Obviously, these are among the knottiest problems around, and a good deal of additional study will be needed before solutions are available. Such study certainly ought to be undertaken, but in view of the complexity of the issues involved, we cannot expect it to produce the desired changes soon.

4. Last, but certainly not least, the Hart-McGovern approach must be considered. Having been critical of all the alternatives presented thus far, it is with pleasure that we turn to one proposal that appears to show a good deal of promise. This is the proposal to enlarge the rights of citizens to review agency determinations in court. The theory behind it is rather simple. If an agency is insensitive to the interests of citizens because of some of the factors mentioned earlier, those citizens ought to have recourse to another, perhaps more objective, forum. This is the theory implicit in the bill entitled the Environmental Protection Act of 1971, which was introduced in Congress by Senators Philip Hart and George McGovern. It is based upon the same theory that is inherent in a similar bill that became law in the State of Michigan in 1970 under the guidance of Professor Joseph Sax of the University of Michigan.

What the Michigan law does and what the Hart–McGovern bill would do, if it became law, is to recognize the right of all citizens to a decent environment and to allow them to enforce that right by enjoining in a court of law all 'unreasonable pollution, impairment, and destruction' of their air, water, and land. Among those who could be sued under this approach are agencies who have permitted such unreasonable activity to take place. Citizens would also be able to sue private individuals who behave unreasonably, even if such behavior is countenanced by regulatory agencies concerned with this matter.

The basic argument in support of this approach is that citizens ought to be empowered to defend the environment, since it is their environment that is at stake. Because of his personal interest in the outcome of environmental litigation, the citizen may well prove a better advocate than any government agency authorized to represent him. Where the citizen's and the agency's views conflict, it is felt that the citizen should be entitled to a complete review of that agency's activities before the judicial branch of government.

This is not to say that the citizen has no right to review agency activities under existing law. He does. That right, however, is at present severely circumscribed. In most cases the citizen will not be able to set aside an agency determination in court unless that determination is illegal, i.e., violates some procedural rule, or is so horrendous as to be labelled by a court of law as 'arbitrary and capricious.' The major thrust of the Hart-McGovern bill is that it would remove this restriction on the court, thus enabling the court to question agency determinations of fact and the exercise of agency discretion. Decisions that are not arbitrary, but nonetheless wrong, would thereby be subject to citizen attack.

Many arguments have been advanced against this approach. The bill has been regarded as a radical — some have said dangerous — departure from traditional practice. The argument most often put forward against the bill is that it would open the courts to 'floodgates of litigation'. Yet the Michigan experience has been otherwise. After eight months' experience with the new law, only a dozen or so cases have been brought to court. Moreover, virtually all of these cases have been responsible suits, thus countering the contention that the bill would lead to frequent harassment of the agencies by cranks.

A second and frequently voiced criticism of the bill is that it would 'tie up the works,' or to put it more extremely, lead to the 'destruction of the administrative system.' The Michigan experience would also seem to answer this objection. A total of twelve suits in eight months hardly indicates that administrative agencies have been paralyzed or even unduly restrained in Michigan.

The Michigan experience would also seem to confirm a fact of life, namely that litigation is never undertaken lightly. It is both costly and time consuming, and neither the Hart-McGovern bill nor the Michigan law purports to change that or to give the plaintiff the right to monetary relief. The range of plaintiffs will therefore include only those willing to risk a substantial loss of money (legal fees, court costs, etc.) with no possibility of monetary gain. Although by joining in a class action the plaintiffs can minimize their risks by splitting up their costs, it seems safe to assume that few will go forward.

Many of the important environmental suits thus far have been brought by public-interest law firms who do not charge fees for their services. Were these firms to expand significantly in number, the 'floodgates' and other related arguments might indeed have some force. This does not seem likely, however, since even existing public-interest law firms operate under financial restraints, and unless there is a rather significant change of heart with regard to foundation funds, the floodgates are virtually certain to remain closed.

A distinct cause for concern, put forward by the President's Council on Environmental Quality, is that judges are incompetent to review administrative decisions in the environmental area. This area is so technical, so the argument goes, that judges should refrain from contesting administrators with greater technical expertise. Yet judges have been getting into technical matters for ages. Familiar examples are nuisance cases, malpractice suits, and actions for patent infringement. While it is true that they have little expertise in environmental protection — generally even less than congressmen — the judicial adversary process, unlike the legislative process, is conducive to responsible action in the area. The objective, fact-finding capability of the judicial system, together with its capacity to clarify points of conflict, gives to courts a considerable advantage in dealing with specific environmental issues. It should further be noted that the Hart-McGovern bill would not require courts to set aside administrative determinations when the judges themselves regard the matter as beyond their own competence. It is most likely that courts will be quite reluctant to get involved in complicated factual issues where administrators have already acted. Where they are willing, however, there seems every reason to permit them to do so.

Another argument occasionally heard against this approach is that it would create uncertainty as to what must be done to comply with pollution laws. Admittedly the standard of 'unreasonable pollution' is susceptible to many interpretations. Moreover, since administrative determinations may be set aside, even the individual who follows the regulators' advice may be sued by those dissatisfied with the regulators. If he chooses to comply with all environmental protection standards, he is nonetheless vulnerable, should those standards be found to be inadequate.

This point would be well taken but for the fact that similar uncertainty already exists under present law. In light of prevailing concern for the environment and new technological advances, environmental protection standards are consistently being revised upward. In his environmental message to Congress at the beginning of 1971, the President suggested that this practice would continue. It follows that mere compliance with currently applicable standards now gives industrial defendants little more security than would be allowed them under the Hart-McGovern proposal.

A final argument, and perhaps the best one against the bill, is that it will not provide much of an improvement over existing mechanisms of environmental control. Accepting the contention that administrators often have been unresponsive and biased, we must ask why we should expect judges to do any better. One answer is that in most cases they will not have to. Underlying the bill is the assumption that, whereas the possibility of a law suit will provide a healthy discipline for agencies, very few cases will actually be brought to court. Secondly, it does seem safe to assume that judges, on the whole, will be less susceptible to outside pressure than administrators. In theory at least, judges are wholly separate and distinct from the executive branch of government, even though that branch may have appointed them. In practice, it seems difficult to envision long lunches and gifts flowing from regulated industries to the judiciary.

To summarize, while the benefits offered by the Hart-McGovern approach may turn out to be limited, a good case can be made for trying it out at this time. In our view, it would provide a useful supplement to Dr. Epstein's proposal for increased citizen participation in the administrative process.

The prospect of citizens actively functioning in the agencies and the courts is an appealing, if somewhat uncertain one. To frustrate this participation in deference to the arguments discussed above would be a serious misallocation of priorities — a triumph of administrative convenience over considerations far more important for society. To promote it, on the other hand, would, we submit, benefit significantly both the citizens involved and the environment they are determined to protect.

Toward Action

SIGMUND KVALØY

The Mardøla Waterfall Development

This is a report on the threat by a certain type of industrial development to Norwegian river systems and waterfalls. It is also an account of a method applied to stop the hydroelectric development of the Mardal Waterfall in Eikesdal in west central Norway. Presumably, this method also has relevance to the fight for a healthy environment in other countries.

These problems, not uniquely Norwegian, are vivid examples of the universal mechanism of growing technocracy and industrialization that threatens our environment in a way that is very difficult to stop. Furthermore, I believe that we have, in this country, some rivers and associated ecosystems that should be regarded as invaluable sources of esthetic experience, health, and ecological knowledge for every human being and that Norwegian authorities should be pushed by international organizations into realizing that they have a responsibility to the rest of the world to save some of these treasures of nature, just as the USA has an international responsibility to save the last redwoods. It is my hope that one such push can be made by the Society for Social Responsibility in Science.

The driving force behind the high and accelerating tempo in hydroelectric developments in Norway has been power-consuming metallurgic raw materials industry, often owned and controlled by international corporations. The criticism of Norwegians concerned with the environment has been leveled especially at the aluminum industry, which gained a foothold in Norway after World War II, when the country had to rebuild industry and our leading politicians solved this problem by offering inexpensive hydroelectric power. This may have been a help at that time to some rural districts as each plant offered a large number of jobs. This is no longer the case, since international competition and automation has resulted in a severe reduction of employment in local production processes. Control from abroad has also increased, since Norway is dependent on the goodwill of the aluminum corporations for imports of the necessary raw materials (bauxite). Many other factors also contribute to a situation where the production of raw aluminum is no more a sound part of the Norwegian national economy. But now that the corporations have gained such a strong foothold within our general economy they expand almost at will. Those who really suffer from this situation are the people living in hundreds of small valleys.

Norway was originally settled along the slender green pastures beside the rivers of her narrow valleys. and Norwegian tradition and culture is still attached to these rural societies, where rivers and waterfalls play an integral part. Only few of the old communities benefit directly through worthwhile industrial employment because the power consuming industries are isolated. Norway's most important natural resource is thus utilized in only a few centralized spots, contributing to the break-up of hundreds of communities. Because uprooted people rush to the cities, where they must crowd together, environments are created where pollution and physical stress cannot be controlled by foresight and planning. The extreme speed of the change, from rural to urban living, makes reasonable action virtually impossible.

People who are torn away from their traditional homes and ways of living in this way, suffer a severe loss of identity. This makes it still easier for the urban industrial interests to use them and to move them from one 'rationalized' plant to another as the needs of industry dictate. People with torn roots, told repeatedly that cars and movies are the most important values in life, are not too likely to make themselves felt as an independent political force. I am merely touching on the way the dominating trend in Western economy and culture is transforming the world.

Ecologically speaking, a complex and well-established web of life and culture is being broken down into a simple standardized network. Through this, the global life complex becomes more vulnerable. Without going further into the different arguments needed to defend such a hypothesis scientifically, I will refer to the case of Mardøla, which is as good an example of the problem as can be presented.

Besides the loss suffered by the international community in losing one of the world's largest and most beautiful waterfalls and a unique ecosystem, there are two rural communities which, by stages, are being destroyed by the central authorities and by the aluminum industry which influences the former. The environment that is being destroyed here is one that anyone would want to have his children grow up in. I appeal to all to go there and see for themselves!

The diversion of the Mardøla River is the third major diversion of water from the communities of Eikesdal and Eresfjord for aluminum-producing purposes. Every time diversion was planned, the people of the valley protested vigorously. Norwegians concerned with the environment have placed every conceivable argument before the decision-makers. Nothing has helped.

At the second diversion — in 1958 — the Norwegian River and Electricity Board (NVE) sent a letter to the State Department of Industry, stating that a third diversion — that of the Mardøla River — ought not to be carried out, as the damage would be greater than the gain. The second diversion drastically reduced the volume of the life-giving water that reached the valley, but the farmers were confident that they would be allowed to keep their last river, the Mardøla.

In 1970, when the decision was made by the Norwegian Parliament to divert the Mardøla to another valley — this being considered a better site for the power plant — it was said that the old Electricity Board of 1958 could not bind the new Board of 1970. But not one word was said about the *reasons* for the statement about damage made in 1958.

As no conventional means of communications had helped, the S.N.M. regarded the parliamentary decision of June 1970 as the go-ahead for a nonviolent, direct action of the Gandhian type.

The idea was to make the experiment of going through with a *strictly* Gandhian action for *environmental* protection. As far as we knew, this had not been done before. At any rate, we did not find any reference in the literature regarding 'Gandhi for environment and natural beauty.' But Gandhi is well suited to such action. He experimented with his methods to be a true reflection of his basic intuition: that Man is unbreakably tied to every other creature in a great web of life, and that to inflict damage on your fellows — human or non-human — will damage yourself because you are part of this greater Self. It is tempting to translate this into ecological terms and speak about energy flow, complexity as system protection, and feedback. What ecological insight needs to get effective on the nature-defense front is an emotional engagement. The vivid feeling that *you* are part of the balanced complexity, that when it is hurt, you are hurt. Our way of getting near this was Ecology plus Gandhi.

'The Mardøla Action' took the material form of a tent camp across the advance line for road construction machinery in the Eikesdal Mountains, barring the way to the heart of the river's watershed and the sites where tunnels were to be started for the river diversion.

We tried to build up a society, defending this green spot in a wild mountain area as one home environment, depending on few and simple means, using, but not damaging, nature. We wanted to demonstrate a better future society to make our action a *positive and constructive* one, in keeping with classical Gandhianism.

To a large extent, I feel that we succeeded in establishing such a

291

society. Farmers and city people of extremely varied backgrounds (altogether about 500 persons) worked and lived together in a harmony not easily imagined. The threat of the police and of an anti-demonstration staged by business and industry interests in the nearest town only toughened the ring of loyalty and strengthened the spirit of unbudging non-violence. The direct action lasted for five weeks, after which time most Norwegians had been drawn into hot discussions on environmental politics. One hundred and twenty-eight people were fined, but everyone is set on going into the courts before paying these fines. Court actions have started, and our supreme court is now being drawn into the matter.

One major result of the Mardøla action is the inspiration it evidently gave to protect a lot of environments that were threatened in one way or another. Groups that had accepted destruction almost as a natural law, are now trying new means of breaking through to protect their own vital interests and that of their children. A new Mardøla action is spoken of in connection with the planned giant airfield near Oslo, in connection with new motor ways through suburbs and across our most fertile farm land, pollution by old and new industry, etc. The Lapps of North Norway are planning a massive direct action if nothing else can stop the hydroelectric development of the rivers and lakes of their reindeer district.

Our effort to make the Norwegian Parliament reconsider the Mardøla decision has not, however, succeeded. Prestige and money spent are here at play. We would appreciate very much appeals from abroad to influence such reconsideration. I speak more than anything else on behalf of the farmers of Eikesdal and Eresfjord and their children, to whom they would like to hand over a good and secure home environment. I am trying to speak on behalf of farmers everywhere, to defend the point that rich possibilities should be saved for those who would like to choose a diverse life of work in close connection with Nature. It is also a plea for *controlled* urbanization and for effective democracy.

LEIV KREYBERG

A Proposal

There is little doubt among reasonably well-informed people that our environment has been grossly and dangerously altered in the recent years. Many of these changes are immediately apparent on the basis of simple observation. Everyone has experienced the irritant effect on eyes and ear passages caused by the polluted air in the larger, and even many smaller cities. Everybody has suffered from headaches. And nearly everyone has seen leaves falling in the vicinity of certain factories, has seen rivers depopulated of fish, and lakes rendered unfit for bathing by an overgrowth of algae.

The possible dangers of many of these changes to physical health are receiving increasing attention. Less attention has been paid to the more elusive effect on emotional well-being and a sense of human dignity of the highly unesthetic accumulation of all kinds of garbage all around us, in the fields and in the woods, and on our beaches. I consider the esthetic aspects of pollution not at all negligible. The marble churches and monuments of our cultural heritage are eroded by sulfur fumes, and the tapestries in our museums are ruined. Who is to say whether the damage to the physical health of priest and wardens, or to the condemned cathedrals and the ancient tapestries in our museums is more destructive — or more significant?

It would appear that the desire of the industrialists for continuous growth in production, and of the public for increasing amounts of consumer goods must be paid for by an increasing amount of rubbish, partly indestructible, and partly dangerous. A balance has to be found between reasonable demands for the production of consumer goods, and reasonable demands for protection against the accumulation of rubbish and particularly dangerous rubbish. What do I mean by 'reasonable'? Obviously, reasonable means that decisions must be based on the exercise of reason. And in order to use reason, the availability of information is an essential prerequisite. Many people have an uneasy feeling about some of the changes taking place but lack sufficiently exact and reliable information on the basis of which they could make the necessary decisions. It has been shown that there is sufficient knowledge available on certain subjects for immediate action to be taken, provided there is public demand for such action. However, a great deal of important information is either unknown to, or, worse, deliberately withheld from those charged with making decisions and the general public which should demand these decisions from the politicians.

My first plea, therefore, is for the establishment of an International Environmental Pollution Reference Center, which would serve as a central repository for all known information on this subject. Appropriately catalogued and cross-referenced data would be made available by the Center not only to investigators, but also to the public and political leaders. Out of the very process of recording the existing body of knowledge must emerge the identification of important gaps in our present knowledge, which has been described as a 'sea of ignorance.'

My second plea, therefore, is for the Center to be charged with the responsibility of initiating, either directly or indirectly, the most important research projects to fill in the most urgent gaps in our knowledge. Only when we can say with reasonable certainty which substances in what doses, are dangerous to which organisms and objects, can specific demands for actions be reasonably presented to politicians and political institutions. As long as we talk only in generalities, specific demands can be dismissed easily by pressure-groups which make it their business to be well-informed and are interested in continuing their activities without interference. Rhetoric is not enough. Only well-substantiated claims are likely to be acted upon. And we have very little time!

Toward a Philosophy

ARNE NAESS

The Place of Joy in a World of Fact

'In a socialist society' says Marx in a very famous passage, 'the "frag-mentary man" will be replaced by the "completely developed individual" ... Men would fish, hunt, or engage in literary criticism without be-coming professional fishermen, hunters, or critics.'

Note that two-thirds of the examples of what we shall do under ideal conditions presuppose a wonderful solution of the main environmental problems.

Let us, as an illustration, suppose a certain portion of the population of London, say five million people, one Thursday morning joyfully de-clares 'Let us go hunting before lunch!' It is not difficult to see that the hundred years since Marx have not mainly been used to get nearer the realization of his admirable communist society.

The complete individual is not a specialist; he is a generalist and an amateur. This does not mean that he has no special interests, that he does not sometimes work hard, that he does not partake in arduous tasks from time to time. But he does so from personal inclination, with joy, and within the famework of his value priorities. Marx does not mean that professionals and specialists need be exterminated. But under the present conditions in highly industrialized, urbanized, and cen-tralized societies, it seems to pay to be a professional and a specialist. That is, high social status and 'success' are assured, if one develops to an extreme degree one's specialized know-how. This holds whether your boss is Nixon, Breshnev, or Mao. In the future ideal society, whether outlined by Marx or by more bourgeois prophets, there will be people who might use most of their energy doing highly specialized, difficult things, but as amateurs, that is from a mature philosophy of life and, therefore, from inclination and with joy. In this sense, and only in this sense, will there be no fragmentary men.

So much for utopias. They all presuppose either the nonexistence of environmental problems, or their solution. My next concern is how to get nearer to our utopias.

There tends to be a contradiction between the life and the teachings of

297

fighters for a better environment. But the main avenue of influence in matters of life-style is that of example, not of preaching. So, my conclusion is that pollution problems would not have reached such tremendous proportions if those who experience intensive environmental joys would make those joys manifest in their own lives. This implies sometimes saying 'No, thank you' to high-salary positions; it implies sometimes being fired rather than complying with orders from superiors. But it also implies disappearances from offices and urban areas for longer periods than conventionally accepted, and saying 'No' to urgent appeals for more lectures on pollution.

I suppose we all admire the pioneers who — through endless meetings in contaminated city air — succeeded in establishing wilderness areas in the United States of America. But their constant work in offices and corridors largely ruined their capacity to enjoy the areas. They lost their capacity to show, in *action,* what they cared for. If they really cared for the wilderness, they would much more often *live* in the wilderness. So many overstressed, successful people admire the wilderness only in words. They do not give up their jobs or step down from their exalted positions as chairmen of this or that, as senior members of law firms, or as professors of wilderness, in order to enjoy the areas at least part of the year. If they had done that, thousands might have followed them, and not only a faint trickle.

The common man shows a good portion of skepticism toward verbally expressed valuations that do not color the life style of the propagandist. The environmentalist sometimes succumbs to a joyless life that belies his concern for a better environment.

The cult of dissatisfaction is apt to add to the already fairly advanced joylessness we can find among socially responsible, successful people, and to undermine one of the chief presuppositions of the ecological movement: joy related to environment and nature.

I know that many have turned their backs on more lucrative careers and on a life of security cultivating well-established sciences. But this is not enough. One's life should manifest the peaks of one's value priorities. The *work* for a better environment is after all only of instrumental value; that is, it does not manifest what you consider non-instrumental, intrinsic values. You remain on the level of techniques. But what criterion do we have to follow the lead of personal priorities? We have one that is dangerously underrated among conscientious, responsible people: joy.

Suppose somebody openly adhered to the doctrine that there cannot under any circumstances — funereal or not — be too much cheerfulness. The sad truth is, I think, that he or she would be classified as shallow, cynical, disrespectful, irreligious, mocking.

Søren Kierkegaard is an important figure here. He *seems* to take anguish, desperation, sense of guilt, suffering as the necessary, and sometimes even sufficient, condition of authentic living. But he also insists upon joy as a condition. If done without joy, it does not matter what you do. 'At 70,000 fathoms you should be glad. At that depth even, you should be able to enjoy going to the circus or the botanical gardens with the children on Sundays. At 70,000 fathoms, one should even retain *det glade Sind* (the joyful mind).' He sometimes calls himself *Hilarius,* the one permeated with *hilaritas. Hilaritas* is the Latin word for cheer, from the Greek, *hilarós.*

Let us listen to another great prophet of our time. 'Dread' is the technical existentialist word for the kind of anxiety that opens one's mind to a deeper understanding of life. But it is, according to Heidegger, the philosopher called the 'chief' of modern pessimism, not an isolated sensation of negative color. It is a *nüchterne Angst,* and joy — *die gerüstete Freude* — is somehow there at the same time. It seems to be a complex state of mind, and without joy there is no dread. That is, the one who thinks he has the dread experience, but lacks joy, suffers from an illusion. The dread has an internal relation to joy.

The unwholesome is not only that we are disintegrated, i.e. immature and therefore joyless, but that we glorify immaturity. Do the most influential philosophers of our time and culture represent peaks of maturity or integration? I have not only Heidegger, Sartre, Kierkegaard, Wittgenstein, but also Marx and Nietzsche in mind. Tentatively, I must answer: 'NO.' There are lesser known, but perhaps more mature philosophers, for example Jaspers and Whitehead. But their influence is negligible; they are not glorified, are not the object of any cult.

Should the world's misery, should the approaching ecocatastrophe, make one sad? It seems that as long as the world's population is more than a million, there must at any time of the day and night be at least one human being in utter despair or excruciating pain. And I suppose one might add that each of us who eats at least once a day with (shameful?) appetite, could easily do a little more to relieve the extreme suffering of others. That one does not do so should, perhaps, be a source of uneasiness or even a feeling of guilt.

It is my point that, nevertheless, there is no good reason why we *should feel sad* about all this. According to the philosophies I am defending, it is a sign of immaturity — the immaturity of unconquered passiveness, of disintegration.

The remedy, the psychotherapy against sadness caused by the world's misery, is to do something (however modest in scale) about that misery. I shall refrain from mentioning Florence Nightingale, but let me note that Gandhi loved to care for, wash, massage, etc. lepers; he simply enjoyed it. It is very common to find those who have to do with extreme misery more than usually cheerful.

According to Spinoza, the power of an individual is infinitely small compared with that of the universe, so we must not expect to save the world at large. The main point, and that is built into the basic conceptual framework of Spinoza's philosophy, is that of being active. By interaction with the extreme miseries, cheerfulness is gained. And the interaction *need* not be a direct one. Most of us can do more in indirect ways using our privileged positions in rich societies. If we are not disintegrated, what we do is intrinsically joined with joy.

Behind the prevalent passivity there is a lot of despair, a lot of pessimism about our own capacity to have a good time, to enjoy ourselves, *except* during 'vacations' — that is, during vacuous times — in a private world of thoughtlessness, well isolated from the great issues.

One of the strangest, one of the next to paradoxical theses of Spinoza, and of Thomas Aquinas and many others, is that the so-called knowledge of evil, knowledge of misery, is inadequate knowledge. In short, that there *is* no such phenomenon, whereas there *is* something good to know. Why? Because evil is always an absence of something, a lack of something positive. Spinoza's theory of knowledge is that objects of knowledge are always *something*. Just as when you say that you *see* that the glass is transparent, what you see is, for instance, a red nose *behind* the glass. You do not *see* the transparency, which is no object.

Well, I do not think that the positive nonexistence of evil can be shown without a large amount of redefining of words, but I do not consider the view totally ridiculous. As with so many other strange points in major philosophies, it has an appeal, it points in the right direction without perhaps stating anything clearly in the scientific sense.

Some of my colleagues ask: 'But how can you, as a university professor, teach joy?' It is not difficult at all.

Spinoza operates with three main concepts of joy and three of sorrow: *Laetitia, hilaritas,* and *titillatio* are the three Latin terms for the positive emotions. Translations are, to a surprising degree, arbitrary, because their function, in Spinoza's system, is only revealed when one studies the complex total structure of his system. The isolation of one concept from the others is not possible. The whole is more than the sum of its parts. You must take the system, or leave it, *as a whole.*

The term *laetitia* I translate as 'joy.' It is a generic term comprising several important subgroups of joy. The main classification of joy, in general, is into *hilaritas* (cheerfulness) and *titillatio* (pleasurable excitement). *Hilaritas* is the serene thing, coloring the whole personality, or better, the whole world. It is the joy of the integrated personality acting according to his own value priorities.

300

Joy is not a sensation. It is not something purely subjective. Neither is it located in the brain. Joy is essentially 'joyfulness of things.' Trees may be joyful. In our subjective-inclined culture we would say they are joyful *looking* when we ourselves look at them with joy. But this leads to the notorious distinction between the thing in itself and what we as subjects (arbitrarily) put into it — an influential distinction mainly of profit to those who find that environmentalists occupy themselves with *mere* feelings and subjective, changeable valuation, whereas they, the technocrats, deal with the *real* thing, the atoms, the electric currents, the quantitative and therefore real properties of the world.

Hilaritas is defined by Spinoza as a joy to which every part of the body contributes. It does not affect only a subgroup of functions of the organism, we would say, but every one, and therefore the totality. There cannot be too much of *hilaritas*, Spinoza contends.

The other main kind of joy, *titillatio*, affects a subgroup of the parts of the body. If narrowly based and strong, it dominates and thereby inhibits other joys. There can accordingly be too much of it. Spinoza mentions love of money, sexual infatuation, ambition, and other sources of joy, which are all good in moderate portions. In moderation they do not hamper and inhibit other sources of joy. I should think that the joy derived from doing a job satisfactorily is good, if moderate.

A second classification of joys is that joy from contemplation of ourselves, of our own achievement, creativity, or, more broadly, activeness, and joy from contemplation of causes of joy outside us. The first he calls satisfaction or repose in ourselves, *acquiescentia in se ipso*, the other *amor*. There can be too much of them, however, because they sometimes refer to parts, not the whole.

What refers to the whole of the body, incidentally, refers also to the whole of the conscious mind, according to Spinoza, and to the whole of the universe, or, more generally, to the whole of Nature as far as we interact with it. This can be understood from Spinoza's so-called philosophy of identity, proclaiming ultimate identity of thought and matter, and from his theory of knowledge, which relates all our knowledge of the world to interaction with the body, exactly as biologists tend to do today.

Lack of *acquiescentia in se ipso* accounts for much of the passivity in environmental conflicts shown by an important sector of the public. People in this group are all on the right side. Yet they are wrong because they do not stand up, clearly, and emphatically, and tell how they feel about the pollution in their neighborhoods. They do not tell in public meetings how they, as private persons, feel about it. They do not have sufficient self-respect, respect for their own feelings, faith in their own importance, to stand up. They think they are 'nobodies.' What we

301

ask is *not* that they should fight for change for the better, but just this, that they make crystal clear how they see things. A small minority will then fight — with joy — supported by the outspoken many.

The distinction between pervasive joy, covering all, and the partial joy, need not be considered an absolute dichotomy, but can be graded. The joy may be more or less pervasive. Clearly, the higher degrees of joy require high degrees of integration of the personality, and high degrees of such integration require intense cultivation of the personal aspect of interaction with the environment. It requires a firm grasp of what we call value priorities, but which Spinoza would rather call reality priorities, because of his resolute location of values among 'objective' realities. Spinoza distinguishes degrees of realness and *perfectio*. This term has to do with *facere,* to do, and may also be translated as 'wholeness' or as 'completed.' The perfected is the complete.

Integration of personality presupposes that we never act as functionaries, never act in the mere capacity of specialist, but always as whole personalities conscious of our value priorities and of the need to make manifest those priorities in social, direct action.

The specific thing to be learned from Spinoza and certain modern psychologists is, however, to integrate the value priorities themselves back into the world. We would tend to say 'the world of facts', but the cleavage of value from facts is mainly due to an overestimation of certain scientific traditions from Galilei, confusing the *instrumental* excellence of mechanistic thinking with its properties as a whole philosophy. Spinoza was heavily influenced by mechanical models of matter, but did not extend them to cover 'reality.' His reality was not mechanical, neither value-neutral, nor value-vacuous.

The cleavage into two worlds, the world of fact and the world of values, can theoretically be overcome by placing, as Spinoza does, the joys and other so-called subjective phenomena into the unified total *field* of realities. But this is too much to go into here. I am more concerned with the place of joy among the total experiences. The objectivist conception of value is important, however, in any discussion in which technocrats tend to dismiss cheerfulness of environment as something *'merely* subjective.'

Spinoza makes use of the following short, crisp, and paradoxical definition of joy (*laetitia*): 'Joy is man's transition from greater to lesser perfection.' Somewhat less categorically he sometimes says that joy is the affect by which or through which we make the transition to greater perfection. Instead of perfection we may say integrity or wholeness.

I consider of central importance the difference between these formulations and the subjectivistic ones, proclaiming that joy only *follows* or *accompanies* transitions. The relation between joy and increase of perfection is, according to Spinoza, an *intrinsic* one, it is not extrinsic or external. That is, the two cannot be separated in practice, only conceptually. Such a realist view of joy suggests that joyfulness like color attaches and forms part of objects, but, of course, changes with the medium and must be defined in terms of interaction with organisms.

Joy is linked intrinsically to increase of perfection, and increase of perfection is linked to a dozen other increases: increase in power and virtue, increase in freedom and rationality, increase of activeness, increase in the degree to which we are causes of our own actions. increase in the degree to which our actions are understandable by reference to ourselves, etc.

Joy is thus, in Spinoza's thinking, part of the basic conceptual structure.

The increase in power is an increase in carrying out what we sincerely strive to do. The Latin term *potentia,* from the verb *possum,* I can, does not presuppose that we coerce somebody else. A tyrant may be less powerful than somebody sitting in his prison. This concept has a long tradition and should not be forgotten.

Joy of work, like any other partial joy, can dominate and subdue other sources of joy to such an extent that the result is stagnation or even decrease in power. In the terminology of Spinoza this means a loss of perfection or integration and increased difficulty in reaching the state of *hilaritas,* cheerfulness.

'To be glad' often seems to be equated with enjoying oneself, laughing, relaxing in the sense of being passive. But enjoying oneself by intoxication, which decreases the higher integrations of the nervous system, means resignation. It means giving up the possibility of joyfulness of the whole person. Cheerfulness in the sense of Spinoza may not be expressed in laughter or smiling, but in concentration, present-ness, activeness.

Buddha was an active person, but had more of *acquiescentia in se ipso,* repose in himself. He is said to have reached *nirvana* long before he died. Properly interpreted within Mahayana Buddhism, this involves supreme integration and liberation of the personality, implied by bliss, or, in the terminology of Spinoza, *hilaritas.* The research of Stcherbatsky and others concerning the term *duhkha,* conventionally translated as 'pain', shows that 'pessimistic Buddhism' also has a doctrine of joy as the central aspect of reaching freedom in *nirvana.*

Loosely, one may say that what we are now lacking in our technological age is the repose in oneself. The conditions of life prevent the full development of that self-respect and self-esteem required to reach a stable high degree of *acquiescentia in se ipso.* The term 'alienation', incidentally, is related to the opposite of *in se,* namely *in alio.* We re-

303

pose in something else, something outside achievement in the eyes of others, other directedness, to use a term from sociology.

Humility *(humilitas)*, as defined by Spinoza, is sorrow *(tristitia)* from man's contemplation of his own impotence, his weakness and helplessness. A sorrow always involves decrease of perfection, virtue, freedom. We can get to know adequately more potent things than ourselves — this gives joy — because of our activeness in the very process of knowing it. The realization of our own potency and the active relation to the more potent gives joy. Thus, instead of humility, which is a kind of sorrow, we have three joys. Firstly, there is the joy due to contemplation of our own power, however small. It gives us *acquiescentia in se ipso,* self-respect and contentedness. Next is the joy due to increasing personal, active knowledge of the greater object, for instance the international pollution situation. And thirdly there is the joy due to the (active) interaction, which strictly speaking defines ourselves as well as the objects as fragments of the total field: nature, in the terminology of Spinoza, or the *ksetra* (field) in the terminology of the *Bhagavadita.*

Adequate knowledge always has a joyful personal aspect because it reveals a power, never a weakness, in our personality. In the words of Spinoza himself:

> Therefore, if man, when he contemplates himself, perceives some kind of his impotency, it does not come from this, that he understands himself, but of this, that his power of action is being reduced. To the extent that man knows himself with true rationality *(vera ratione),* to that extent it is assumed that it is his essence he understands, that is his power.

We say, with some haughtiness, that Spinoza belongs to the age of rationalism, to the pre-Freud, pre-Hitler era. But Spinoza in many ways anticipated Freud, and his term *ratio* must not be translated by our 'rational' or 'rationality' without immediately adding that his *ratio* was of a subtle kind and internally related to emotion. Spinoza was not an intellectual. Rational action was to him action with the absolutely maximal perspective, things seen as fragments of total nature, not what we tend to call rational today, not what we in Norway call *snusfornuftig.*

Pity and commiseration *(commiseratio* and *misericordia)* are no virtues with Spinoza, and even less with Gandhi. But there are worse feelings, and pity may have positive instrumental value. Spinoza says:

> . . . commiseration, like shame, although it is not a virtue, is never-

theless good in so far as it shows that a desire of living honestly
(*honeste*) is present in the man who is possessed with shame, just
as pain is called good in so far as it shows that the injured part has
not yet putrified.

A modest function, but nevertheless of instrumental value! Tersely,
Spinoza adds that 'a man lives according to the dictates of reason,
strives as much as possible to prevent himself from being touched by
commiseration.'

Commiseration is sorrow, and therefore is in itself an evil. According
to a certain conventional morality, a duty should be carried out, even
if there is no joy. The above suggests that we had better *disregard our
duties, if not permeated with joy.* This seems to me to be somewhat
fanatical, except when adding a kind of norm concerning the high
priority of development of the *capacity* of joy. 'Alas! I cannot do my
duty today because it does not fill me with joy. I had better escalate
my efforts to experience joy!' Spinoza does not stress the remedy,
integration. But he does presuppose it.

The case of humility shows how *ratio* changes sorrows to joys:
Spinozistic psychoanalysis tries to loosen up mental cramps causing
unnecessary pain.

Freud worked with a tripartition, the *Id (das Es)*, the *Ego*, and the
Superego. The Superego has, through its main application in explanation of neuroses, acquired a rather ugly reputation; it coerces the poor
individual to try the impossible and lets him then experience shame
and humility when there is no success.

In Spinoza's analysis, the *ratio* also functions as a kind of overseer,
but its main function is rather one of consolation. It fixes the attention
upon what we can rather than upon what we cannot, and eliminates
feelings of necessary separation from others, stressing the harmony of
rational wills, and of well understood self-interests.

A major virtue of a system like Spinoza's is the extreme consistency
and tenacity with which consequences, even the most paradoxical, are
drawn from intuitively reasonable principles. It meets the requirements
of clarity and logic of modern natural science. The system says, for
instance, to us: 'You do not like consequence No. 101? But you admit
it follows from a premise you have admitted. Then give up the premise.'
Or: 'You do not want to give up the premise? Then you must give
up the logic, the rules of inference, the rules you use to deduce the
conclusion. You cannot give them up, you say? But then you have to
accept the consequences, namely the logical conclusion. You don't want
to do that either? Well, I suppose you don't want clarity and integration
of your views and your personality.' The rationality of a total view
like Spinoza's is perhaps the only form of rationality able to break down

the pseudo-rational thinking of the conservative technocracy that now obstructs the effort to think in terms of the total biosphere and its development in the near and more remote future.

There are powerful premises represented in Chinese, Indian, Islamic, Hebrew, as well as in Western philosophy, namely those that have so-called *ultimate unity of all life* as a slogan. They do not hide the fact that big fish eat small ones, but they stress the profound interdependence and functional unity of such a biospheric magnitude that non-violence, mutual respect, feelings of identity are always potentially there, even between the killer and the so-called victim.

In many cultures the identification is not limited to life, but envelops the mineral world as well, making us conceive ourselves as genuine surface fragments of our planet, rather special fragments, capable of somehow experiencing the existence of other fragments. A microcosm in macrocosm.

Another idea right at the basis of a system from which the environmental norms are derivable is that of *self-realization. The mature human* individual, with a widened self, acknowledges a right to self-realization that is universal and seeks a social order, or rather a biospherical order, that maximizes the potential of self-realization of all kinds of beings.

Level-headed and tough-minded environmentalists sometimes stress that it is sheer hypocrisy to pretend that we try to protect nature. In reality, they say, we always have the needs of human beings in view. This is false, I think. Thousands of supporters of unpolluted so-called wastelands in Northern Labrador simply wish that those lands *should be there* as they are. For their own sake. They are of so-called intrinsic, not only instrumental value. To invoke *specifically* human needs here is misleading, just as it is misleading to say that it is egoistic to share one's birthday cake with others because one *likes* to share with others.

Self-realization is not maximal realization of the coercive powers of the ego. The self in the kinds of philosophy I am alluding to is something expansive, and the environmental crisis may turn out to be of immense value for the further expansion of human consciousness.

In modern education the difference between a world picture (or better, a world model) and a straightforward description of the world is slurred over. The atoms, particles, wave functions are presented as part or fragments of nature, even as *the real,* objective nature, as contrasted with the human projections into nature: the 'colorful' but 'subjective' nature.

But physical reality, in terms of modern science, is perhaps only a piece of abstract mathematical reality — a reality we emphatically do not *live* in. Let us use π as an illustration. If some of our mathematical colleagues are not badly mistaken, then somewhere far out in the infinite decimal expansion of π, all our telephone numbers appear, one after another, arranged alphabetically, and there appear, somewhere else, one million zeros followed by the number of our hairs minus three. Let us agree that these fragments of the unending decimal series do exist, but let us also agree that the term 'exist' does not have much to do with our living environment. That environment is made up of all the colorful, odorous, ugly, and beautiful details, and it is sheer folly to look for an existing thing without color, odor, or any other homely qualities.

Philosophically, there are great problems in this area. The main one is not too difficult to formulate: Take a walk in winter with one hand in your pocket and the other exposed to the wind. Put both hands in water when you get back home.

It happens that according to one of the hands the water *is* cold, according to the other hand the water *is* warm. Broadly speaking the Western philosophical tradition favors the view that *in reality* the water is neither cold nor warm, these qualities somehow being created in our brains or somehow projected into the water. The Eastern tradition tends more in the direction of saying: both! That is, the water is both warm and cold, the cold hand perceiving the warmth, the warm hand getting to know the coldness.

Contradiction? Can the water have both qualities? Yes, because the negation of 'x is warm' is 'It is not so that x is warm.' And the negation of 'x is cold' is 'It is not so that x is cold.' The two sentences 'x is cold' and 'It is not so that x is warm' are *not* synonymous. There is not, therefore, a logical contradiction in holding them both to be true. It only makes it necessary to modify our ways of conceiving the relation between reality and appearance.

I think this *both*-cold-*and*-warm theory is the best proposed so far. It destroys the usual classroom view, namely that the so-called primary sense qualities, movements, shape, solidity, etc., are real, objective qualities, whereas colors, odors, etc., the secondary and tertiary qualities, are subjective, somehow created in (so far) utterly inexplicable ways, or by means of certain cellular processes in the brain.

The significance of this subject is a broadly cultural one: the rehabilitation of *die bunte Welt* — the gaudy world — the status of the immediately experienced world, the colorful and joyful world.

What is the answer to the question *Where* is joy in the world of fact? It can be only one: 'Right at the Center'.

About the authors

Ralph Nader is Director of the Centre for the Study of Responsive Law in Washington, D. C.

Dr. Alice Mary Hilton is a Consultant for Systems Analysis and Computing Machines and Professor of Mathematical Logic at Queen's College, University of the City of New York. She is a past President of SSRS.

Professor M. W. Thring is Head of the Department of Mechanical Engineering at Queen Mary College of the University of London.

Professor Widukind Lenz is Director of the Institute of Human Genetics at the University of Münster. He is the recipient of the Albert Einstein Award of the Society for Social Responsibility in Science.

Ruth M. Harmer is the author of several books on pollution and the environment.

Ernest J. Sternglass is Professor of Radiation Physics and Director of Radiological Physics in the Department of Radiology, School of Public Health, University of Pittsburg.

Dr. E. L. Bourodimos is Professor of Civil and Environmental Engineering at Rutger's University, New Jersey.

Dr. Björn O. Gillberg is with the Department of Microbiology of the Royal Agricultural College of Sweden in Uppsala.

Øistein Strømnæs is Assistant Professor at the Institute of General Genetics at the University of Oslo.

Agnar P. Nygaard is Professor of Biochemistry at the University of Bergen.

William C. Davidon is Professor of Physics at Heverford College, Pennsylvania.

Victor Paschkis is Professor Emeritus of Columbia University of New York. He is one of the founders and a past President of SSRS.

Dr. Lars Emmelin is with the Environmental Studies Programme of the University of Lund.

Professor Wouter van Ramsdonk is on the faculty of the Vrije Universiteit in Amsterdam.

Harry Briggs is the Editor and Publisher of *Time for Answers*.

Leonard Bickwit is Staff Counsel of the (U.S.) Senate Committee on Commerce in Washington, D. C.

Michael B. Brownlee is on the staff of the (U.S.) Senate Committee on Commerce in Washington, D. C.

Sigmund Kvaløy is Lecturer in Philosophy at the University of Oslo.

Leiv Kreyberg was Professor of Pathology at the University of Oslo from 1938 to 1964.

Arne Næss was Professor of Philosophy at the University of Oslo from 1939 to 1970.